THE **COMPLETE**
IDIOT'S
GUIDE TO

Algebra Word Problems

Wauconda Area Library
801 N. Main Street
Wauconda, IL 60084

Algebra Word Problems

Wauconda Area Library
801 N. Main Street
Wauconda, IL 60084

by Izolda Fotiyeva, Ph.D.

ALPHA
A member of Penguin Group (USA) Inc.

ALPHA BOOKS

Published by the Penguin Group

Penguin Group (USA) Inc., 375 Hudson Street, New York, New York 10014, USA

Penguin Group (Canada), 90 Eglinton Avenue East, Suite 700, Toronto, Ontario M4P 2Y3, Canada (a division of Pearson Penguin Canada Inc.)

Penguin Books Ltd., 80 Strand, London WC2R 0RL, England

Penguin Ireland, 25 St. Stephen's Green, Dublin 2, Ireland (a division of Penguin Books Ltd.)

Penguin Group (Australia), 250 Camberwell Road, Camberwell, Victoria 3124, Australia (a division of Pearson Australia Group Pty. Ltd.)

Penguin Books India Pvt. Ltd., 11 Community Centre, Panchsheel Park, New Delhi—110 017, India

Penguin Group (NZ), 67 Apollo Drive, Rosedale, North Shore, Auckland 1311, New Zealand (a division of Pearson New Zealand Ltd.)

Penguin Books (South Africa) (Pty.) Ltd., 24 Sturdee Avenue, Rosebank, Johannesburg 2196, South Africa

Penguin Books Ltd., Registered Offices: 80 Strand, London WC2R 0RL, England

International Standard Book Number: 978-1-61564-037-9
Library of Congress Catalog Card Number: 2010902235

12   11   10   8   7   6   5   4   3   2   1

Interpretation of the printing code: The rightmost number of the first series of numbers is the year of the book's printing; the rightmost number of the second series of numbers is the number of the book's printing. For example, a printing code of 10-1 shows that the first printing occurred in 2010.

Printed in the United States of America

Note: This publication contains the opinions and ideas of its author. It is intended to provide helpful and informative material on the subject matter covered. It is sold with the understanding that the author and publisher are not engaged in rendering professional services in the book. If the reader requires personal assistance or advice, a competent professional should be consulted.

The author and publisher specifically disclaim any responsibility for any liability, loss, or risk, personal or otherwise, which is incurred as a consequence, directly or indirectly, of the use and application of any of the contents of this book.

Most Alpha books are available at special quantity discounts for bulk purchases for sales promotions, premiums, fund-raising, or educational use. Special books, or book excerpts, can also be created to fit specific needs.

For details, write: Special Markets, Alpha Books, 375 Hudson Street, New York, NY 10014.

Publisher: *Marie Butler-Knight*
Associate Publisher: *Mike Sanders*
Senior Managing Editor: *Billy Fields*
Senior Acquisitions Editor: *Paul Dinas*
Development Editor: *Nancy D. Lewis*
Production Editor: *Kayla Dugger*

Copy Editor: *Krista Hansing Editorial Services, Inc.*
Cover Designer: *William Thomas*
Book Designers: *William Thomas, Rebecca Batchelor*
Indexer: *Julie Bess*
Layout: *Rebecca Batchelor*
Proofreader: *John Etchison*

I dedicate this book to my wonderful sons Dmitriy and Michael Fotiyev. Being your mom is the best thing that has ever happened to me. And to my extended family for their love and support.

Contents

Introduction

I've written this book so that you can master many different techniques and apply them to any algebra word problem you encounter in the future. All concepts in this book are carefully explained, important definitions and procedures are provided, and many sample problems with step-by-step solutions appear in every chapter.

This book has 20 chapters. The first four chapters refresh your knowledge of algebra concepts, rules, and methods of solving linear equations and systems of linear equations. Starting with Chapter 5, each chapter deals with a specific type of algebra word problem: problems involving proportions, age problems, and motion problems, to name just a few.

Every chapter opens with an "In This Chapter" list that presents some of the major concepts and topics covered. All concepts pertaining to a specific type of word problem are thoroughly explained; all definitions, formulas, and procedures are clearly presented. Each concept and formula is reinforced by clearly written sample problems and accompanied by step-by-step solutions.

Beginning with Chapter 5, each chapter contains seven practice problems that cover the important concepts of the chapter. We learn best by doing, and you'll solidify your understanding of the chapter's material by working to solve similar problems right after you study the material. Solve these problems, check your answers, and then compare them with the worked-out solutions in Appendix D. These practice problems will help you determine whether you have mastered the material completely or need further study on the topic.

Finally, each chapter wraps up with a "The Least You Need to Know" section that summarizes the key material and lists the main formulas. Review this section, and you'll know the most important concepts and formulas of each chapter.

As you begin your studies, I want to offer some suggestions for using this book and for achieving your goal to master word problems: Read each chapter with a pencil (or pen) in hand. Move through the sample examples with great care and take your time to fully understand each step of the solution. As you proceed through the book, don't get frustrated if you don't understand something after the first reading. Try again—things will become clearer as you read and try to follow each step several times.

After you learn the material presented in a chapter, solve all the practice problems. The more problems you solve, the better you will understand the material.

How This Book Is Organized

This book is presented in six parts:

In **Part 1, Solving Word Problems,** you review basic rules of algebra, firm up your skills in solving linear equations and systems of linear equations, and equip yourself with general strategies and tips to meet the challenges of algebra word problems.

In **Part 2, Ratio, Proportions, and Percentage Problems,** you get acquainted with ways to write ratios, as well as solve directly and inversely proportional problems. You also learn a percent formula that helps you deal with almost any percent problem.

In **Part 3, Number, Age, and Work Problems,** you discover techniques for finding missing numbers or digits, learn to cope with age problems, and face the so-called "work" problems in which people and/or pipes work together to get the job done faster.

Part 4, Money, Mixtures, and Discount Problems, introduces you to the host of problems that deal with different financial aspects: calculating the amount of money, investing with simple and compound interest, and finding the discount and original prices of items.

In **Part 5, Geometry, Physics, and Liquid Solution Problems,** you learn to write algebraic equations to solve geometry problems, find the ways to express temperature using different scales, and get fully prepared to mix solutions to obtain the desired strength.

Finally, in **Part 6, Average and Motion Problems,** you get a chance to calculate average and weighted average scores, and you come face-to-face with the dreaded motion problems, with things and people moving in the same and opposite directions. You wrap up your study by solving word problems that belong to more than one category.

Extras

You'll find these different sidebars throughout the book:

DEFINITION

Algebra is full of terms, notions, and rules that require an explanation. To master algebra word problems, you need to have full command of them.

A WORD OF ADVICE

Frequently, you will wonder "How did she do this?" or "How did she transform this into that?" This type of sidebar gives you tips and shortcuts to save you a lot of time and eliminate frustration.

DEAD ENDS

Although I caution you against common pitfalls throughout each chapter, these sidebars warn you about the most dangerous ones that are capable of completely throwing you off track.

WORTHY TO KNOW

These little snippets will broaden your view of the topic at hand while providing some useful knowledge, facts, and trivia. They also shed some light on what lies beyond the scope of this book.

Acknowledgments

I would like to take this opportunity to thank many people who made this book possible. Your help, comments, and criticism were extremely helpful.

Very special thanks are due to my literary agent, Marilyn Allen, at Allen O'Shea Literary Agency, who brought this book project into my life.

I am very grateful to the staff of Alpha Books, especially to my acquisitions editor, Paul Dinas, for his valuable advice, direction, encouragement, and contagious laugh that always eased my author's worries. I also want to thank my development editor, Nancy Lewis, for her professionalism and commitment.

All in-chapter illustrations were created by Dmitriy Fotiyev, my terrific son, who is an electrical engineer and computer scientist by training. I also need to thank my younger son, Michael, who is a college student and pretty amazing as well, for reviewing and proofreading the manuscript before it hit the tables of the real editors at Alpha Books.

I am indebted to my brother-in-law, Boris Meyerson, who agreed to technically review this book before I sent it to the editors. Boris earned a Master's degree in electrical engineering from Ural State Polytechnic University in Russia and currently works as a computer scientist. I am very grateful for his help and expertise.

Finally, I need to thank my extended family for their continuous support, love, and belief in me.

Special Thanks to the Technical Reviewer

The Complete Idiot's Guide to Algebra Word Problems was reviewed by an expert who double-checked the accuracy of what you'll learn here, to help us ensure that this book gives you everything you need to know about working algebra word problems. Special thanks are extended to Dawn B. Sova, Ph.D.

Dawn Sova, who earned a doctorate from Drew University in 1990, is the author of *How to Solve Word Problems in Geometry* and *Schaum's Quick Guide to the Verbal SAT*, in addition to 22 books and numerous magazine articles. In 2002, she received the Best Nonfiction Category Award from Mystery Writers of America for her 2001 work *Edgar Allen Poe, A to Z*; she had been nominated in the same category in 1997 for her book *Agatha Christie A to Z*. At the college level, Dr. Sova has taught a range of writing courses, including magazine writing, journalism, news writing, and investigative reporting. Her book *How to Write Articles for Newspapers and Magazines* is used as a supplemental textbook in many colleges. As a high school instructor, she developed the SAT Review curriculum for both the Verbal and Math sections and has served as a consultant to school districts in course development. She has also worked with the Educational Testing Service (ETS) as a grader for SAT and GRE essays, and for Advanced Placement exams. The author currently serves as a curriculum consultant and faculty course mentor for Thomas Edison State College in the Directed Independent Adult Learning (D.I.A.L.) program and oversees Portfolio Learning Assessment for several writing courses at the college.

Trademarks

Solving Word Problems

We begin our study of algebra word problems with a short review of algebra itself: its terminology, rules, and properties. Keep in mind that algebraic concepts are developed as an extension of familiar arithmetic ideas and rules. Your knowledge of arithmetic will help you master algebra concepts and apply them in solving algebra word problems.

In this part, we also review methods of solving linear equations and systems of linear equations so that you're fully equipped to deal with different word problems.

Who's Afraid of Word Problems?

In This Chapter

- Suggestions for solving word problems
- Determining a problem's question and eliminating the unneeded information
- Choosing a variable wisely
- Using the "looking back" approach

Algebra word problems consist of words—no guessing here. Apart from pure algebraic problems that are completely abstract and don't relate to the real world, algebra word problems represent that same real world. That is why they are sometimes called real-world problems. The range of topics is endless: the interest earned on invested money, unknown digits in a number, concentration of a solution, round-trip travel, and much more.

Algebra word problems are notorious for being confusing, difficult, and incomprehensible. In this chapter, we discuss the steps to master general strategies and techniques that help us "attack" any algebra word problem.

What's Algebra Got to Do with It?

Many word problems can be solved with the help of arithmetic; however, there are many more that arithmetic is powerless to deal with. This is where algebra comes to the rescue. In this book, we discuss problems that can be solved by using either algebraic linear equations or systems of linear equations with two variables.

Many people become discouraged and frustrated when they first encounter algebra word problems. Problem solving is not something that you can master overnight. It takes a lot of hard work, patience, perseverance, and time. If you persist and try to solve many problems, gradually you will become more confident with them.

No single procedure or formula will ensure success at solving algebra word problems, but the following suggestions and steps can be effective and helpful.

Step 1: Read the problem slowly and carefully, making certain that you understand the meaning of all the words and technical terms. For instance, if the problem deals with two consecutive even numbers, you must know the meaning of this technical phrase.

Step 2: Read the problem a second time (perhaps even a third time) to get a clear overview of the problem's situation and determine the known facts. Try to state the problem using your own words, and restate what is given and what is to be found.

Step 3: Try to identify the type of the problem at hand and recollect the general techniques and formulas to be used for this particular type of word problem.

Step 4: Determine the problem's question. The question is usually located at the end of the problem. The keywords to look for are *how far, how many, how much time,* and similar ones. As soon as you identify the problem's question, you will have more direction in how to solve the problem.

Step 5: Choose a meaningful variable to represent an unknown quantity in the problem and express any other unknowns in terms of that variable.

Step 6: Create a sketch, diagram, table, or chart that might help you visualize the problem and understand it better.

Step 7: Set up an equation that contains the variable and the problem's known facts by translating the conditions of the problem from English into algebraic language.

Step 8: Solve the equation and use the solution to compute all facts requested in the problem.

Step 9: Check your answers.

In the sections that follow, we discuss more thoroughly some of the steps we've just mentioned.

How Much Information Do We Really Need?

Some people complain that word problems have … too many words in them. The issue with this complaint is that, along with some numbers, each word problem contains words that are crucial for solving the problem. Other words are just the ballast that makes the problem look like a neat story from the real world. These words can be the names of people or cities, the makeup of cars, the brand names of food or clothing people buy, and so forth.

A WORD OF ADVICE

Formulas are always a good starting point in solving word problems. Memorize them or at least keep them handy for quick reference. Make sure that you are familiar with what the variables in each formula stand for.

To illustrate this point, consider a problem about two sisters: Anna and Amy leave their house at 10 A.M. and start walking in opposite directions. Anna took 30 minutes to browse at Payless Shoes Store, where she bought two pairs of shoes. Amy met her friend Allie and chatted with her for 40 minutes; then they had lunch together, and it was Amy's treat. Both sisters spend $98 together. If Anna spent $7 less than twice the amount of money that Amy spent on lunch, how much money did each of the sisters spend this morning?

The problem mentions that the sisters started walking in opposite directions at 10 A.M. It also includes times that the sisters spend on their activities listed. We really don't need all these pieces of information, since the problem's question is about the money that each of them spent this morning. So just ignore this information and concentrate on the facts that help to answer the problem's question.

If we try to eliminate all unneeded information, the problem looks similar to this: Two sisters, Anna and Amy, spent $98 dollars together. If Anna spent $7 less than twice the amount of money that Amy spent, how much money did each of the sisters spend?

Certainly, the initial problem is an extreme example of unneeded information. In reality, only one or two facts from a certain problem, if any, will be redundant or unnecessary. Still, you must learn how to choose the information that is relevant to solving a problem and ignore the rest.

Furthermore, soon you will realize that many algebra word problems may not seem practical or even realistic. Don't allow this fact to distract you. Keep in mind that the real purpose of these problems is not to reflect real-life situations accurately, but to allow you to develop problem-solving techniques.

The last point I want to make here is this: as you read the problem carefully and try to eliminate the unneeded information and keep the important pieces, try to simultaneously decide which type of word problem you are dealing with.

Straightening Up the Units

In this book, you will encounter many algebra word problems that involve units of time, length, weight, area, and money. Some of them will involve two or more units—such as minutes and hours or inches and feet—and you will need to decide which unit to use.

The choice of units depends on convenience and common sense. If you are measuring the width of a table, you use feet. If you are measuring the distance between two towns, miles is the appropriate unit to use. If the measure of a table is given in both feet and inches, you have two choices: to use only feet or to use only inches. For instance, if the length of a table is 3 feet and 8 inches, you can convert feet into inches using the fact that 1 feet = 12 inches. 3 feet and 8 inches = $(3 \cdot 12) + 8 = 36 + 8 = 44$ inches. Another possibility here is to use fractions and convert 8 inches into feet: $\frac{8}{12} = \frac{2}{3}$ feet, so the length of the table is $3\frac{2}{3}$ feet. The choice is yours; the only advice here is to change everything to the same units before you do computations and set up an equation.

As the book proceeds with various types of word problems, I discuss in more detail how to convert units and how to decide which measures to use.

Visualizing the Word Problem

When solving word problems, it is helpful to imagine yourself actually in the word problem and try to visualize the problem's situation. Then reread the problem several times until you have a clear picture in your mind.

As you visualize the problem, you should start organizing it by creating a table, a chart, or a sketch. Remember that even the simplest drawing can help you understand the problem better. For example, if six people bought two tickets each for $5 a ticket, the drawing could be a stick figure holding two tickets with "6 people, 2 tickets, $5" written underneath.

How can you know what to use to organize a problem: a figure, a chart, or a diagram? As we progress through the book, you'll learn how to organize each type of word problem—hopefully, by the end of the book, you will be an ace in terms of visualizing and organizing a problem. We consider some general tips on this topic for now.

Use the following problem as an example:

> Nick is three times younger than his mom, and his sister is 20 years younger than the mom. Their dad is 5 years older than the mom. Find the age of each family member if their combined age is 145 years.

Let's discuss the choices we have for organizing the problem, keeping in mind that whatever we create should simplify the problem and suggest how to proceed with its solution.

A drawing including all the people and their ages will be too confusing and cumbersome. A figure that we usually create for geometry problems is out of the question as well. A chart or a table definitely is the best way to organize this problem.

Next, we need to decide how many rows and columns our chart should have. It is customary to list people, types of products, or mixtures in the first column. Since we are dealing with four people in this problem (Nick, his sister, the mom, and the dad), the first column in our chart will have the initial four rows. Also, we must label our columns, so all together we should have five rows. In addition to people, the other crucial information to display is the people's corresponding age, which we list in the second column. We are ready to create our chart:

People	Age
Nick	3 times younger than his mom
His sister	20 years younger than the mom
The mom	?
The dad	5 years older than the mom

A good table or chart must contain all the important information from the problem and be organized in a clear and concise way.

As we proceed through the book, tables and charts will contain more columns and more information. Moreover, before filling in our tables and charts, we will need to do some preliminary computations, since the problems will not always provide us with all the values and facts so readily. But don't worry about this now; when the time comes, you will learn this step by step, and you will have enough opportunities to master the art of a problem's organization.

> **WORTHY TO KNOW**
>
> Examples of word problems can be found dating back to Babylonian and Egyptian times. In many Babylonian problems, students had to find the lengths of canals dug, weights of stones, lengths of broken reeds, areas of fields, numbers of bricks used in a construction, and so on. The famous Rhind Mathematical Papyrus, which is now at the British Museum, is the best example of Egyptian mathematics and includes some interesting word problems.

For geometry problems, creating figures and sketches is great: they help in writing expressions about the relationships between angles, sides, and areas. Many motion and lever problems will benefit greatly from a good sketch. But even a good drawing loses all its power if the labels are omitted. By labeling the different parts of your drawings—angle measures, rates in miles per hours, distances in miles, and so forth—you increase your chances of writing the right equation for the problem.

Sometimes the problem already includes a sketch. It is still helpful to redraw the sketch and show all the important facts.

Hopefully by the end of the book, creating visual aids and organizing word problems in a meaningful way will become second nature to you.

Solving the Problem with an Equation

Another important step in solving algebra word problems is to choose a meaningful variable to represent an unknown quantity and to express other values in terms of this variable. Let's discuss this by using the same problem about the family members and their ages that we used earlier.

Notice that in the preceding table, we have a question mark for the mom's age, so it is tempting to represent the mom's age by the variable x. Let's do this and see if it works. We declare that the mom's age is x. Next, we go through the table and try to express other values using the same variable. We know that Nick is three times younger than his mom, so his age can be expressed by dividing his mom's age by 3: $\frac{x}{3}$. Now let's stop and think. Since we expressed Nick's age as a fraction, our equation most certainly will have fractions in it, and fractions are much more difficult to work with than whole numbers. Can we do something about this? Yes, we can.

To say that Nick is three times younger than his mom is the same as saying that the mom is three times older than Nick. We can then represent Nick's age with the variable x, and the mom's age becomes $3x$ (three times her son's age). By doing this, we forced the fraction out of our equation and made it easier to solve. This little exercise proves that our task in solving word problems is not only to choose the variable, but most important, to choose it wisely.

Let's continue to express other facts from the problem in terms of the variable. The dad's age becomes $3x + 5$, since he is 5 years older than the mom, and the sister's age is $3x - 20$, since she is 20 years younger than the mom. Now we can plug the found variables and values into our table:

People	Age
Nick	x
His sister	$3x - 20$
The mom	$3x$
The dad	$3x + 5$

After we choose a variable wisely, express the problem's other facts in terms of the variable, and organize the problem with a chart or a sketch, we must look for a guideline that can be helpful in setting up a problem's equation.

A WORD OF ADVICE

Often formulas are used within the solution of a problem but do not serve as the main guideline for setting up the problem's equation. For instance, motion problems always involve the main formula $d = r \cdot t$, but the main guideline for such problems is usually a statement about either distances, times, or speeds (rates). Be careful in identifying the right guideline.

A guideline might be a statement of a relationship, such as "The difference of the two numbers is 63." Or a guideline might be a formula, such as "Distance equals rate multiplied by time." In the case of our problem, the guideline is the statement that the combined family age is 145 years. Since we know all the ages expressed in terms of the variable x, we can add them all up and make the sum equal to 145 years: $x + (3x - 20) + 3x + (3x + 5) = 145$.

We will not proceed further with solving the equation, since we need to refresh our memory about algebra rules and methods of solving equations in the following chapters.

Checking the Answer

After you solve the problem's equation and find the solutions, plug all answers back into the original statement of the problem. If you decide not to check an answer, at least check that the answer is reasonable enough. For example, if the problem involves two investments totaling $20,000, then an answer of $25,000 for one investment is not reasonable at all.

Another important facet of word problem solving is looking back over the solution and reflecting on the following questions:

1. Do you see another plan that could be used to solve the problem? Is there a shorter way to find the solution or another guideline that could be used?

2. Is this particular problem closely related to another problem that you have recently solved?

3. Have you "stored" the reasoning and techniques used to solve this problem for future reference?

The approach of looking back over the solution and reflecting on it is a powerful way to lay a foundation for solving future algebra word problems.

We discussed the important facets of problem solving in general. The next step is solving various types of algebra word problems and learning specific strategies and approaches for each type.

But before we start sorting word problems by types or categories and solving them, let's delve into the realm of algebra and refresh our memory about the major terms, rules, and operations of this mathematical discipline.

The Least You Need to Know

- Reading a word problem several times and retelling it in your own words ensures a full understanding of the problem's situation.
- Sketching a figure, chart, or diagram is helpful in analyzing the problem.
- Using a guideline helps to set up a problem's equation.
- Checking your answers is a good way to avoid mistakes.

Let's Talk Algebra

In This Chapter

- Getting familiar with algebra terms
- Classifying types of numbers
- Using the order of operations
- Learning to deal with signs

The word *algebra* originated from the Arabic word *al-jabr*, which first appeared in the book *A Handbook of Calculations by Completion and Reduction*, written by Middle Eastern astronomer and mathematician al-Khawarizmi (787–850). This was the most significant textbook in the field of algebra written since the time of the Greek mathematician Diophantus, who lived 600 years earlier. Diophantus knew how to solve many equations—he even introduced and widely used literal symbolism—but the Arabs invented the real methods of solving equations.

The Arab mathematicians knew how Babylonian and Indian mathematicians worked with equations, and they borrowed their knowledge of equations, at the same time significantly improving their method of solving problems. But their method had one major drawback: when writing equations, they didn't use either letters or numbers—they used only words. You can imagine how cumbersome and inconvenient this way of writing equations was. For instance, to write down the short equation like $x^2 + 3x = 28$, the Arabic method used more than 30 words!

In the twelfth century, the book was translated into Latin, the official language of science in medieval Europe. The translator didn't translate the word *al-jabr*; he simply wrote it down using Latin letters as *algebra*. When European mathematicians mastered al-Khawarizmi's methods of equation solving, they started to improve them and apply them to more difficult equations. Gradually, letter symbolism became inevitable as equations became more complicated. During the fifteenth to seventeenth centuries, algebra became a legitimate branch of mathematics, and mathematicians began solving equations in much the same way we do so today. Many outstanding mathematicians contributed to the development of algebra, including René Descartes, François Vieta, and Gottfried Wilhelm Leibniz, who once exclaimed, "People who are not familiar with algebra can't imagine all wonderful things they might have achieved with the help of this science."

In this chapter, we review some algebraic terminology and rules, and explore how to use the language of algebra.

WORTHY TO KNOW

Interestingly, the word *al-jabr* was widely used nonmathematically in Europe. It became known through the Moors of Spain, where an *algebrista* is a bonesetter, or "restorer of bones." Often a barber of medieval towns called himself an algebrista, since barbers often performed bloodletting and bonesetting on the side. One such algebrista is featured in the famous book *Don Quixote,* by Spanish writer Miguel de Cervantes (1547–1616).

Algebra Terminology

Learning algebra resembles learning a foreign language: one needs to learn words and rules by which words are connected into sentences. Luckily, algebra is a relatively simple language; it doesn't have hundreds of thousands of words to memorize like, for instance, the English language. Algebra uses the letters of the Latin alphabet—a, b, c ... x, y, z—to write its "words." Moreover, algebra uses the same symbols as arithmetic to indicate addition, subtraction, multiplication, and division. That means we are already familiar with the basic vocabulary. But we still need to learn new vocabulary words to master the "language" of algebra.

The most important term in algebra is a *variable*. We use variables when we need to show how objects are related to each other. Objects can be anything, from the pairs of shoes we buy to the number of bacteria growing inside a rotting apple. For example, if we want to say that the length of a rectangle is 5 times the length of its width, we can write this relation as $l = 5w$. We call l and w variables because the numbers they represent can vary—that is, we can substitute many pairs of numbers in the expression to satisfy it. For instance, w could be 5, so the corresponding l would then be 25; or w could be 4, so then l would become 20, and so forth.

Sometimes we encounter variables standing alone and connected by signs of operations, as in $x + y$. More often, however, variables are accompanied by a number, as in $2x + 3y$. We call these numbers *coefficients*. Strictly speaking, the coefficients in front of the variables in $x + y$ are 1 and 1, but it is customary to omit the 1 as a coefficient in such cases.

Constants are parts of algebraic phrases and contain only numbers. Constants don't have variable parts; they never change their value. In the expression $8xy - 3x + 7$, 7 is a constant.

DEFINITION

A **variable** is a symbol assigned to an unknown value. Letters such as a, b, x, y, z, n, r, d, and t usually represent a variable. A **coefficient** is the number that appears at the beginning of a monomial; for example, the coefficient of $13yz$ is 13. A **constant** or a constant term is a monomial without a variable. In $3xy + 5$, the number 5 is the constant term.

Terms are building blocks of any algebraic phrase. A term can be any of the following:

- A constant: $2, -\dfrac{3}{4}, 43$

- A product of a coefficient and a variable: $10x, 3y$

- A product of two or more variables: xy^2, a^4b, xyz

Terms are called like terms when they have the same variable part and differ only in their numerical coefficients. For instance, terms $5xy^2$ and $16xy^2$ are like terms since they have the same variable part. We collect or combine like terms by adding their numerical coefficients ($5 + 16 = 21$) and keeping the same variable part. In our example, the result of collecting is one term, $21xy^2$.

DEAD ENDS

All constants are like terms as well, so you need to combine them. In the following expression $4xy + 2 + 3xy - 5 + 8$, the free terms or constants 2, –5, and 8 are all one set of like terms, and the other set is $4xy$ and $3xy$. To combine constants is pretty easy—you just add them together: $2 + (-5) + 8 = 5$. Then collect the terms with variables: $4xy + 3xy = 7xy$. Thus, the original expression is equal to $7xy + 5$.

Terms such as $5xy$ and $7xy^2$ are not like terms, since their variable parts are different (even if they share the same variable x). But the exponents for variable y are different. Hence, we can't call these terms like terms and can't combine them. They should stay separate.

An algebraic expression is one or more terms connected by signs of operations. The following are examples of algebraic expressions: $3a - 4b$, $xy + 3xz^2$, $5x^2 - 6x + 3$.

Two algebraic expressions separated by an equals sign form an equation. The algebraic expression on the left side has the same value as the expression on the right side:

$4x + 3 = 2x + 9$

When a variable in an equation is replaced by a number, the result may be true or false. If the result is true, then the number is called a solution of the equation. We also can say that the number satisfies the equation. For example, if we substitute 2 for x in the previous equation, the left side of the equation is no longer equal to the right side:

$4(2) + 3 \neq 2(2) + 9$

$11 \neq 13$

On the other side, if we plug in 3 for x in the same equation, both sides of the equation are still equal:

$4(3) + 3 = 2(3) + 9$

$15 = 15$

We can state that 3 is the solution of the equation.

Algebra deals with the set of real numbers that comprise many kinds of numbers: natural, integers, fractions, decimals, rational, and irrational numbers. Let's discuss each type briefly:

- Natural numbers (the counting numbers): 1, 2, 3, 4, 5 …

- Whole numbers: 0, 1, 2, 3, 4, 5 …

- Integers (whole numbers and their opposites): … –2, –1, 0, 1, 2 …

- Rational numbers: They can be expressed as a ratio of two nonzero integers $\frac{m}{n}$. The decimal equivalent of a rational number can be a terminating decimal (ends somewhere). For example, $\frac{2}{5} = 0.4$ and $\frac{1}{8} = 0.125$. The decimal equivalent can also be a repeating decimal. This is a decimal that has an infinitely repeating sequence of digits. For example, $\frac{1}{3} = 0.33333... = 0.\bar{3}$. The horizontal bar over 3 lets us know that 3 repeats and never stops. Natural numbers, integers, fractions, and terminating and repeating decimals are examples of rational numbers.

- Irrational numbers are numbers that can't be expressed as a ratio of two integers. The decimal equivalent of an irrational number is a never-terminating and never-repeating decimal: 0.1342563 ... or 1.0200653034 ... Examples of irrational numbers are $\sqrt{2}$, $\sqrt{7}$, $\sqrt{14}$, and so forth.

WORTHY TO KNOW

The Greek philosopher Pythagoras founded a scholarly community in Southern Italy in 539 B.C.E.—a secret society somewhere between a religious order and a university. The Pythagoreans were the members of this society. They all took oaths to ensure that their discoveries remained with the Pythagorean society. Since they were so secretive, we know very little about them.

It is believed that one of them, Hippasus, proved the existence of irrational numbers at a time when the Pythagorean belief was that whole numbers and their ratios could describe anything that was geometric. Not only that, they didn't believe there was a need for any other numbers. According to one legend, Hippasus was subsequently thrown overboard by his fellows or died in a shipwreck for revealing the truth to nonmembers of the society. Another legend states that Hippasus was merely exiled for this revelation. Thus, it is true that the Pythagoreans discovered irrational numbers; everything else is a combination of reality and legends.

Which numbers are there more of, rational or irrational? The answer to this question at first seems obvious: rational numbers—after all, they include an infinite number of natural numbers, all integers, fractions, and many decimals, and we deal mostly with them in our daily lives. But that's false: there are many more irrational numbers than rational numbers. This fact has been proven mathematically—there are small islands of rational numbers floating in the huge ocean of irrationality!

As we stated before, rational and irrational numbers together comprise the set of real numbers. Real numbers can be modeled as points on a horizontal line called a real number line. The point for number 0 is the origin. Negative numbers are represented by points to the left of 0, and positive numbers are represented by points to the right of 0.

The Real Number Line.

Basic Rules of Algebra

When we drive a car, fly an aircraft, or operate machinery, we need to follow some specific rules. Otherwise, we either get into an accident or break whatever we are navigating or operating at the moment. Before taking on any task, we'd better learn some basic rules and follow them. The same goes with algebra: before writing and solving equations for word problems, we need to master some important algebra rules.

Order of Operations

For algebraic expressions that have more than one operation, it is important to know which operation to perform first. In algebra, first priority is given to exponents; then we need to take care of multiplication and division. Addition and subtraction have last priority.

If a problem includes only operations that have the same priority, they are performed from left to right. For example:

$12 + 14 − 5 = (12 + 14) − 5 = 26 − 5 = 21$

$32 : 8 \cdot 3 = (32 : 8) \cdot 3 = 4 \cdot 3 = 12$

In a problem such as this:

$12 + 18 : 3 − 5 = 12 + (6) − 5 = 18 − 5 = 13$

The operations are not performed from left to right, since we need to perform division first as the operation of higher priority and then continue from left to right.

Grouping symbols, such as parentheses () or brackets [], indicate the order in which the operations should be performed. Consider the following example:

$$(2 \cdot 3)^2 + (6 \div 2)^2 = 6^2 + 3^2 = 36 + 9 = 45$$

Parentheses dictate that we first perform the operations of multiplication and division inside parentheses and then take care of exponents.

Without the grouping symbols, the same expression has a different value, since we first need to do exponents:

$$2 \cdot 3^2 + 6 \div 2^2 = 2 \cdot 9 + 6 \div 4 = 18 + 1.5 = 19.5$$

Operations with Real Numbers

We will use the *absolute value* to do operations with real numbers.

DEFINITION

Absolute value is the distance of a number from zero on the number line. It is always positive. $|6| = 6$, $|-6| = 6$

The absolute value of the number x, denoted as $|x|$, can be interpreted geometrically as the distance from 0 to x on the number line. Since absolute value describes a distance, it is never negative. Consider some examples:

$|7| = 7$

$|-12| = 12$

$|7 - 2| = |5| = 5$

$|2 - 13| = |-11| = 11$

When adding two numbers with the same sign, add their absolute values and keep the common sign as the sign of the sum. For example:

$4 + 5 = 9$

$-12 + (-5) = -17$

When adding two numbers with different signs, subtract the smallest absolute value from the greater absolute value, and use the sign of the number with the greater absolute value as the sign for the sum. For example:

$-13 + 17 = +4$

$21 + (-28) = -7$

To subtract signed numbers, change the subtraction operation to addition using the definition of subtraction. Then add, using the rules of addition for signed numbers. For example:

$7 - 10 = 7 + (-10) = -3$

$5 - (-18) = 5 + 18 = 23$

$-21 - (-3) = -21 + 3 = -18$

A WORD OF ADVICE

For two real numbers a and b, $a - b = a + (-b)$. Expressed in words: to subtract b from a, add the additive inverse of b to a. The result is called *the difference.*

When you get used to the operation of subtraction, you will be able to perform the middle step of replacing subtraction with addition in your head.

We can now try to simplify algebraic expressions using our knowledge of operations with signed numbers. Let's try to simplify the expression $-5x - 11y - 3x + 17y$. In this expression, we have two pairs of like terms: one pair with variable x, another with variable y. We will keep the variables but add the coefficients in front of them:

$-5x - 3x = [-5 + (-3)]x = (-8)x = -8x$

$-11y + 17y = (-11 + 17)y = (6)y = 6y$

Then, the whole expression becomes:

$-5x - 11y - 3x + 17y = -8x + 6y$

One last thing that we need to discuss in this section is how to multiply and divide signed numbers. The rules are the same for both operations and are not difficult at all: when the two signs are the same, the product or quotient is positive; when the two signs are different, the product or quotient is negative.

For example, for multiplication we have the following:

$(+2) \cdot (+3) = 6$

$(-2) \cdot (-3) = 6$

$(+4) \cdot (-3) = -12$

$(-5) \cdot (+2) = -10$

For division, we have:

$(+6):(+2) = 3$

$(-12):(-3) = 4$

$(-14):(+2) = -7$

$(+20):(-5) = -4$

What if we need to multiply more than two signed numbers? The product of an even number of negative signs is positive. For instance, the following products are positive, since the number of negative signs is even:

$(-2)(-8) = 16$

$(-6)(-2)(-3)(-4) = 144$

The product of an odd number of negative signs is negative. For example, the following products are negative, since the number of negative signs is odd:

$(-5)(-2)(-4) = -40$

$(-2)(-3)(-1)(-4)(-5) = -120$

Properties of Addition and Multiplication

From your past experience of working with whole numbers and fractions, you probably remember their basic properties and rules. Now we extend these properties to include real numbers and algebraic expressions. The following table summarizes these properties. In this table, a, b, and c represent real numbers, variables, or algebraic expressions.

Name	Property	Example
Commutative Property of Addition	$a + b = b + a$	$x^3 + 2 = 2 + x^3$
Associative Property of Addition	$(a + b) + c = a + (b + c)$	$(x^2 + 3x) + y = x^2 + (3x + y)$
Additive Identity Property	$a + 0 = a$	$4y + 0 = 4y$
Additive Inverse Property	$a + (-a) = 0$	$8x^3 + (-8x^3) = 0$
Commutative Property of Multiplication	$a \cdot b = b \cdot a$	$y(x + 3) = (x + 3)y$
Associative Property of Multiplication	$(ab) \cdot c = a(bc)$	$(5x \cdot y)7 = 5x(y \cdot 7)$
Multiplicative Identity Property	$a \cdot 1 = a$	$y \cdot 1 = y$
Multiplicative Inverse Property	$a \cdot \dfrac{1}{a} = 1$	$\left(5 - x^3\right) \cdot \dfrac{1}{\left(5 - x^3\right)} = 1$
Distributive Properties	$a(b + c) = ab + ac$ $(a + b)c = ac + bc$	$4(6 + 7x) = 24 + 28x$ $(y + 10)3 = 3y + 30$
Multiplication Property of -1	$(-1) \cdot a = -a$ $a \cdot (-1) = -a$	$(-1) \cdot x = -x$ $5x \cdot (-1) = -5x$
Double Negative Property	$-(-a) = a$	$-(-2y) = 2y$

Note that distributive properties are also true for subtraction, since subtraction is defined as "adding the opposite."

Review these properties carefully. You will use them extensively while solving equations for word problems.

Don't Get Lost in Translation

When translating verbal phrases into algebraic expressions, we need to look for some important keywords that will help us identify the operation and mathematical phrase.

Keywords that indicate **addition** include: *sum, total, more than, greater than,* and *older than.*

Keywords that indicate **subtraction** include: *difference, less than, minus, subtracted from,* and *younger than.*

Keywords that indicate **multiplication** include: *product, multiplied by, twice, doubled, tripled,* and *times.*

Keywords that indicate **division** are: *quotient, divided by,* and *per.*

Keywords that translate into an **equals sign** are: *is, was, will be, equal,* and *are.*

The examples that follow illustrate how we use some of these keywords to translate verbal phrases into algebraic language:

The sum of two numbers is 65.

$a + b = 65$

The height is 5 more than the length.

$h = l + 5$

Bill is 6 years older than Anna.

$B = A + 6$

A number plus 7 is 13.

$x + 7 = 13$

The number *a* is 6 more than the number *b*.

$a = 6 + b$

One number is $\frac{2}{3}$ of another.

$a = \frac{2}{3}b$

Refer to Appendix B for the fuller list of words, phrases, and their algebraic equivalents.

The Least You Need to Know

- According to the order of operations, the first priority is given to exponents and the last priority is given to addition and subtraction.
- Grouping symbols such as parentheses () or brackets [ ] indicate the order in which the operations should be performed.

- Use the absolute value to do operations with real numbers.
- Properties of addition and multiplication for real numbers and algebraic expressions are an extension of the same properties for whole numbers and fractions.
- Keywords help to identify the operation and translate verbal phrases into algebraic expressions.

Give Those Equations a Makeover

In This Chapter

- Understanding equivalent linear equations
- Dealing with variables on both sides of an equation
- Opening parentheses and collecting like terms
- Finding the LCD in fractional equations

In this chapter, we consider techniques and procedures for solving *linear equations in one variable*. The following are examples of these equations:

$2x + 3 = 7$

$5x + 4x = 8x - 2$

Solving an equation entails finding the number or numbers that make an algebraic equation a true numerical statement, as we discussed in Chapter 2. Remember that such a number is called the solution of the equation.

Balancing Linear Equations

Consider the following three equations:

$7x - 3 = 4x + 12$

$3x = 15$

$x = 5$

We can verify that the number 5 is the solution for all three of them by plugging 5 into each equation and obtaining true numerical statements.

Equations that have the same solution are called equivalent equations. The general procedure for solving an equation is to continue replacing the given equation with equivalent equations that are simpler after each step until the equation has this form: variable = constant or constant = variable.

In the preceding example, the equation $7x - 3 = 4x + 12$ has been first transformed into $3x = 15$, and then further simplified to $x = 5$ (variable = constant).

A WORD OF ADVICE

The final goal in the process of solving linear equations is to get the variable x all by itself on the left (or right) side of the equation.

For the rest of this chapter, our concern is various procedures for simplifying equations and obtaining the form of variable = constant. The process of simplifying an equation and obtaining x all by itself is called isolating the variable. We use this term extensively in future chapters.

Two properties of equality play a crucial role in solving equations. The first of these is called the *addition-subtraction property of equality*.

DEFINITION

Linear equations in one variable contain only one variable which has an exponent of one.

The addition-subtraction property of equality states that any number can be added to or subtracted from both sides of an equation to produce an equivalent equation:

For all real numbers $a, b,$ and $c,$

$a = b$ if and only if $a + c = b + c;$

$a = b$ if and only if $a - c = b - c.$

Sample Problem 1

To isolate the variable (that is, get it all by itself on the left side) in the following equation, we need to make –11 disappear:

$x - 11 = 5$

Well, we're not magicians, so it is beyond our human ability to make it physically disappear. But we have mathematics on our side, so we can use the previously mentioned property to turn –11 into 0 on the left by adding 11 to both sides:

$x - 11 + 11 = 5 + 11$

$x = 5 + 11$

$x = 16$

Solution: $x = 16$

Consider one more example.

Sample Problem 2

$22 = y + 13$

In this case, to isolate the variable y on the right side, we need to subtract 13 from both sides:

$22 - 13 = y + 13 - 13$

$22 - 13 = y$

$9 = y$

We obtained the solution $9 = y$. Using the *symmetry property of equality*, we can change it into $y = 9$.

Solution: $y = 9$

We extensively use the symmetry property of equality in the next chapters, so store it in your memory for future reference.

The second important property pertaining to solving equations is called the *multiplication-division property of equality*.

DEFINITION

The **symmetry property of equality** states: if $a = b$, then $b = a$.

The multiplication-division property of equality states that an equivalent equation is obtained whenever both sides of the original equation are multiplied or divided by the same nonzero real number:

For all real numbers a, b, and c, where $c \neq 0$,

$a = b$ if and only if $ac = bc$;

$a = b$ if and only if $\dfrac{a}{c} = \dfrac{b}{c}$.

We illustrate the use of this property with two examples.

Sample Problem 3

In this example, we need to get rid of the fraction in front of the variable:

$$\frac{5}{6}x = 35$$

To achieve this, we multiply both sides by $\frac{6}{5}$ (which is the reciprocal of $\frac{5}{6}$):

$$\frac{6}{5}\left(\frac{5}{6}\right)x = \frac{6}{5}(35)$$

$$x = 42$$

Solution: $x = 42$

A WORD OF ADVICE

When a fraction is multiplied by its reciprocal, the result is always 1: $\frac{3}{4} \cdot \frac{4}{3} = 1$.

This is based on the multiplicative inverse property from the table in Chapter 2.

Sample Problem 4 (Method 1)

In the following example, to get the variable all by itself on the right side, we need to divide both sides by -5:

$$45 = -5x$$

$$\frac{45}{-5} = \frac{-5x}{-5}$$

$$-9 = x \text{ or } x = -9$$

Solution: $x = -9$

Note that, in this example, we divided both sides by -5 instead of 5, since our goal is to obtain an equivalent equation in the form of variable = constant, where the variable must always be positive.

Looking back at the last two examples, notice that when the coefficient of the variable was an integer, we divided both sides by this integer; when the coefficient of the variable was a fraction, we multiplied by the reciprocal of the fraction. Generally, since dividing by a number is equivalent to multiplying by its reciprocal, the multiplication-division property can be stated only in terms of multiplication.

We illustrate the last point by redoing the previous example with the help of multiplication by the reciprocal.

Sample Problem 4 (Method 2)

$45 = -5x$

Multiply both sides by the reciprocal of -5, which is $-\dfrac{1}{5}$:

$$\left(-\frac{1}{5}\right)45 = \left(-\frac{1}{5}\right)(-5)x$$

$-9 = x$ or $x = -9$

Solution: $x = -9$

We can easily realize that it is still more convenient to divide by the whole number than to multiply by its reciprocal.

Summarizing what has been said about the multiplication-division property of equality, we can state that when the coefficient of x is an integer or a decimal, we should divide both sides of the equation by that number. When the coefficient of x is a fraction, we should multiply both sides of the equation by its reciprocal.

Sometimes the process of transforming a linear equation is compared to the process of balancing a scale. If you remove some weight from one side of a scale, you need to remove the same weight from the other side to keep the scale balanced. If you increase the weight on one side by a factor of two, you need to do the same with the other side. The analogy is helpful to keep your equations balanced.

Transforming Linear Equations

Most of the time, multiple steps are required to solve a linear equation. This happens, for example, when we have variables on both sides. Then, to obtain the equivalent equation in the form variable = constant, we need to apply several transformations to the original equation.

Sample Problem 5

$x + 8 = 2x - 12$

The strategy for solving such an equation is to isolate the variable terms on one side (usually on the left), and then obtain all free terms on the other side and collect like terms, if needed. Let's do this step by step. To isolate the variable terms on the left, we first need to subtract $2x$ from both sides:

$x - 2x + 8 = 2x - 2x - 12$

Collect like terms:

$-x + 8 = -12$

To continue isolating the variable, we need to subtract 8 from both sides:

$-x + 8 - 8 = -12 - 8$

Collect like terms:

$-x = -20$

Divide both sides by -1:

$x = 20$

Solution: $x = 20$

DEAD ENDS

Remember that, technically, a negative variable ($-x$) still has a coefficient of -1, which is usually omitted. It can be written as $-1x$. So to make a variable positive, divide both sides by -1.

After enough practice, you will be able to do at least half of the steps in your head. In the following chapters, I omit some of these steps to make the solutions shorter. For example, the preceding equation will be solved this way:

$x + 8 = 2x - 12$

Isolate the variable:

$x - 2x = -12 - 8$

Collect like terms:

$-x = -20$

Divide both sides by -1:

$x = 20$

Let's solve another equation using all the necessary steps.

Sample Problem 6

$$y + 6 = \frac{1}{8}y + 10$$

To isolate the variable, first subtract $\frac{1}{8}y$, and then subtract 6 from both sides, collecting like terms in the process:

$$y - \frac{1}{8}y + 6 = \frac{1}{8}y - \frac{1}{8}y + 10$$

$$y - \frac{1}{8}y + 6 = 10$$

$$y - \frac{1}{8}y + 6 - 6 = 10 - 6$$

$$y - \frac{1}{8}y = 4$$

Next, we need to collect like terms on the left. To subtract $y - \frac{1}{8}y$, remember that y and $1y$ mean the same thing, so we can rewrite the last expression as $1y - \frac{1}{8}y$. Use the common denominator of 8 to obtain the answer: $\frac{8}{8}y - \frac{1}{8}y = \frac{7}{8}y$. Our transformed equation is now the following:

$$\frac{7}{8}y = 4$$

Finally, since the coefficient in front of y is fractional, in order to obtain y all by itself, we must multiply both sides by the fraction's reciprocal:

$$\frac{8}{7}\left(\frac{7}{8}y\right) = \left(\frac{8}{7}\right)4$$

$$y = \frac{32}{7} = 4\frac{4}{7}$$

Solution: $y = 4\frac{4}{7}$

Simplifying Linear Equations

Having variables on both sides is not the only time we use multiple steps to solve an equation. When the equation involves grouping symbols, the solving process again requires multiple transformations. Let's discuss this with the following example.

Sample Problem 7

$6(2x - 1) - 4(5 - x) = 6$

The first step here is to remove parentheses by applying the distributive property (see the table in Chapter 2):

$12x - 6 - 20 + 4x = 6$

Make sure that, when using the distributive property, you multiply the signed numbers correctly.

Collect like terms on the left:

$16x - 26 = 6$

Isolate the variable term by adding 26 to both sides:

$16x - 26 + 26 = 6 + 26$

Collect like terms:

$16x = 32$

Divide both sides by 16:

$x = 2$

Solution: $x = 2$

Let's solve one more equation with grouping symbols.

Sample Problem 8

$3(x - 1) = 3(3x - 2) + 27$

Distribute on both sides:

$3x - 3 = 9x - 6 + 27$

Collect like terms on the right:

$3x - 3 = 9x + 21$

A WORD OF ADVICE

Before adding or subtracting terms on both sides to isolate the variable, always collect like terms on each side first, if needed.

Subtract $9x$ from both sides and collect like terms:

$3x - 9x - 3 = 9x - 9x + 21$

$-6x - 3 = 21$

Add 3 to both sides and collect like terms:

$-6x - 3 + 3 = 21 + 3$

$-6x = 24$

Divide both sides by -6:

$x = -4$

Solution: $x = -4$

Fractions and Decimals in Linear Equations

Fractions are a headache for so many people. To have fractions in an equation is like having a migraine. But we don't have any other choice than to endure this and discuss it thoroughly, because many word problems you'll deal with in the future will have fractional equations.

Frequently, the fractional coefficient stands in front of a binomial, as in the following example.

Sample Problem 9

$$x - 4 = \frac{1}{3}(x + 32)$$

To get rid of the fraction, multiply both sides by its reciprocal, which is just 3:

$$3(x - 4) = 3 \cdot \frac{1}{3}(x + 32)$$
$$3(x - 4) = (x + 32)$$

Distribute on both sides:

$3x - 12 = x + 32$

Isolate the variable and collect like terms (notice that I am taking a shortcut here, as we will be doing in the future):

$3x - x = 12 + 32$

$2x = 44$

Divide both sides by 2:

$x = 22$

Solution: $x = 22$

When an equation contains fractions, our foremost goal is to form an equivalent equation containing only integers. To accomplish this, the general technique is to multiply both sides of the equation by the *least common denominator* (*LCD*) of all the fractions.

DEFINITION

The **least common denominator (LCD)** is the smallest common multiple of the denominators of two or more fractions. The LCD of $\frac{3}{4}$ and $\frac{2}{5}$ is 20.

Sample Problem 10

$$\frac{3}{4}x - \frac{1}{3}x = 10$$

Multiply both sides of the equation by the LCD, which is 12:

$$12\left(\frac{3}{4}x - \frac{1}{3}x\right) = 12(10)$$

A WORD OF ADVICE

For example, the LCD for 12, 16, and 18 can be obtained as follows:

Factor the integers into their prime factors and write the factors in exponent form:

$12 = 2 \cdot 2 \cdot 3 = 2^2 \cdot 3$

$16 = 2 \cdot 2 \cdot 2 \cdot 2 = 2^4$

$18 = 2 \cdot 3 \cdot 3 = 2 \cdot 3^2$

Take all the bases to their highest exponent and multiply. The product is the LCD. The bases are 2 and 3. The highest exponent of 2 is 4, and that of 3 is 2. Hence, the LCD is $2^4 \cdot 3^2 = 16 \cdot 9 = 144$.

Distribute on the left side, keeping in mind that when we multiply a whole number by a fraction, we need to multiply this number by the fraction's numerator and leave the denominator unchanged:

$$\frac{12 \cdot 3}{4} x - \frac{12 \cdot 1}{3} x = 120$$

Reduce by common factors and multiply the remaining factors:

$9x - 4x = 120$

Collect like terms on the right and then divide both sides by 5:

$5x = 120$

$x = 24$

Solution: $x = 24$

Let's solve one more equation with fractions.

Sample Problem 11

$$\frac{8}{9} x - \frac{1}{6} x = \frac{3}{4} x + \frac{1}{8}$$

First find the LCD:

$9 = 3^2$, $6 = 2 \cdot 3$, $4 = 2^2$, and $8 = 2^3$

LCD is $2^3 \cdot 3^2 = 72$

Instead of multiplying both sides by the LCD and then distributing, we can multiply each term by the LCD (both ways are completely correct):

$$\frac{72 \cdot 8}{9} x - \frac{72 \cdot 1}{6} x = \frac{72 \cdot 3}{4} x + \frac{72 \cdot 1}{8}$$

Reduce by common factors and multiply the remaining factors:

$64x - 12x = 54x + 9$

Collect like terms on the left side:

$52x = 54x + 9$

Isolate the variable and collect like terms:

$52x - 54x = 54x - 54x + 9$

$-2x = 9$

Divide both sides by –2:

$$x = -\frac{9}{2}$$

Solution: $x = -\frac{9}{2}$

Of all the fractional equations we'll solve in this book, one of the most common is the proportion, an equation in which two ratios are set equal to one another, like this: $\frac{a}{b} = \frac{c}{d}$. Let's solve the next equation and discuss a shortcut.

Sample Problem 12

$$\frac{x+2}{5} = \frac{x-1}{2}$$

Instead of finding the LCD of the equation's fractions and proceeding the usual way, we can use the abovementioned shortcut, called *cross-multiplication*. This method eliminates the denominators of the fractions and doesn't require us to find the LCD.

DEFINITION

To use **cross-multiplication,** multiply each numerator by the other side's denominator and set both products equal.

If $\frac{a}{b} = \frac{c}{d}$, then $a \cdot d = c \cdot b$.

Use cross-multiplication:

$(x + 2)2 = (x - 1)5$

Using the commutative property of multiplication, we can rewrite this as:

$2(x + 2) = 5(x - 1)$

Using the symmetry property of equality, we can also write:

$5(x - 1) = 2(x + 2)$

All three of the last equations are the same equation written in a different form. In the future, we will extensively use the two properties that we've just mentioned to write the results of cross-multiplication in the most convenient way.

Let's continue with our solution. We will solve the following form:

$2(x + 2) = 5(x - 1)$

Distribute:

$2x + 4 = 5x - 5$

Isolate the variable and collect like terms:

$2x - 5x = -4 - 5$

$-3x = -9$

Divide by -3:

$x = 3$

Solution: $x = 3$

When an equation has decimals in it, we have two ways to solve it. The first method is to deal with decimals directly and perform all the steps the same way we did with integers. Let's illustrate this with the following example.

Sample Problem 13 (Method 1)

$0.35x + 0.10(30 - x) = (0.20)(30)$

Distribute on the left side and multiply on the right side:

$0.35x + 3 - 0.10x = 6$

Isolate the variable and collect like terms:

$0.35x - 0.10x = 6 - 3$

$0.25x = 3$

Divide both sides by 0.25:

$x = 12$

Solution: $x = 12$

Some people don't feel comfortable with decimals when they need to add them and divide by them (even though they use calculators anyway), so another choice is to multiply both sides by 100 (or by any other multiple of 10) to eliminate decimals. Let's solve the preceding equation using this method.

Sample Problem 13 (Method 2)

$0.35x + 0.10(30 - x) = (0.20)(30)$

Distribute on the left side and multiply on the right side:

$0.35x + 3 - 0.10x = 6$

Multiply both sides by 100:

$100(0.35x + 3 - 0.10x) = 100(6)$

$35x + 300 - 10x = 600$

Isolate the variable and collect like terms:

$35x - 10x = 600 - 300$

$25x = 300$

Divide both sides by 25:

$x = 12$

Solution: $x = 12$

We obtained the same answer, $x = 12$. In the future, we will use both methods.

After this review, hopefully you will be able to solve any linear equation, whether with or without fractions or decimals.

The Least You Need to Know

- The general procedure for solving an equation is to continue replacing the given equation with equivalent equations until an equation has the form variable = constant.
- If an equation contains grouping symbols, open the parentheses before proceeding with the addition-subtraction/multiplication-division properties of equality.
- To eliminate fractions, multiply each term by the LCD and simplify right away. The resulting equation should contain only integer coefficients.
- To eliminate decimals, multiply each term (or each side and then distribute) by a multiple of 10.

Sometimes One Is Just Not Enough

In This Chapter

- Different methods of solving systems of linear equations
- Obtaining a revised equation for the substitution method
- Multiplying to apply the addition method
- Choosing which method to use

So far we have been dealing with linear equations in one variable. If an equation has two variables in it (usually denoted as x and y), it is impossible to solve it using the methods we discussed in Chapter 3. One equation is just not enough to obtain the solution. We need one more equation with the same two variables to find the values of x and y. Even though some new methods will be introduced in this chapter, we will still be using the properties and rules related to solving linear equations.

What Are Systems of Linear Equations?

Our challenge for the rest of this chapter is to learn how to solve *systems of linear equations*. Solving a system of linear equations means finding all the *ordered pairs* that are solutions of all the equations in the system. When we find the *solution set* of each system, we provide our answers as follows: the solution set is $\{(x, y)\}$.

> **DEFINITION**
>
> Two equations in two variables considered together are called a **system of linear equations.**
>
> The **ordered pair** is a pair of real numbers (usually denoted as *x* and *y*) that satisfies all the equations in the system. These numbers are also called the **solution set** of the system. The pair is said to be "ordered" because, when you list the solution, the value for *x* always comes before the value for *y*.

The following are systems of linear equations:

$$\begin{cases} 2x + 3y = 27 \\ x - y = 11 \end{cases}$$

$$\begin{cases} 3a + 4b = 253 \\ 5a - b = 0 \end{cases}$$

Notice that some systems can contain variables other than *x* and *y*. Also note the grouping symbol at the left of the system. This symbol indicates that the two equations should be considered and solved together.

In this chapter, we consider only systems with two linear equations in two variables. Several techniques are used to solve these systems; we use only two in this book (the substitution method and the addition method).

The Substitution Method

The first method of solving systems of linear equations is called the substitution method. It can be used on any system; however, we will see very soon that some systems lend themselves more easily to this method than others.

The key idea behind the substitution method is to eliminate one of the variables. The substitution method is especially helpful if one of the equations is of the form *x* equals or *y* equals.

Let's use the substitution method to solve a few examples and see its benefits and drawbacks.

Sample Problem 1

Solve the following system:

$$\begin{cases} y = 2x + 20 \\ x + y = 14 \end{cases}$$

Since the first equation states that y equals $2x + 20$, we can substitute the expression $2x + 20$ for y in the second equation of the system:

$x + y = 14$

$x + (2x + 20) = 14$

WORTHY TO KNOW

Scientists and engineers use systems of linear equations that consist of hundreds of equations to design electrical circuits and microchips, develop mathematical models in economics, and schedule flight crews, to name just a few applications.

Now we've obtained an equation with one variable that can be solved using the usual methods. Open parentheses and collect like terms on the left:

$x + 2x + 20 = 14$

$3x + 20 = 14$

Isolate the variable and collect like terms on the right:

$3x = -20 + 14$

$3x = -6$

Divide both sides by 3:

$x = -2$

To find the value of y, substitute 2 for x in one of the original equations. I substitute it into the first equation:

$y = 2x + 20 = 2(-2) + 20 = -4 + 20 = 16$

Solution set: $\{(-2, 16)\}$

Sample Problem 2

Let's consider another system:

$$\begin{cases} x = 4y + 10 \\ 3x + 5y = -4 \end{cases}$$

Since the first equation states that x equals $4y + 10$, we can substitute the expression $4y + 10$ for x into the second equation of the system:

$3x + 5y = -4$

$3(4y + 10) + 5y = -4$

We obtained an equation with one variable. Open parentheses and collect like terms on the left:

$12y + 30 + 5y = -4$

$17y + 30 = -4$

Isolate the variable and collect like terms on the right:

$17y = -30 - 4$

$17y = -34$

Divide both sides by 17:

$y = -2$

To find the value of x, substitute -2 for y in one of the original equations. I substitute y into the first equation here:

$x = 4y + 10$

$x = 4(-2) + 10 = = -8 + 10 = 2$

Solution set: $\{(2, -2)\}$

It may be necessary to change the form of one of the system's equations before it is suitable for substitution. Let's illustrate this point with the following example.

Sample Problem 3

$$\begin{cases} 2x + 3y = 27 \\ x - y = 11 \end{cases}$$

None of the equations is ready for substitution right away. We can notice, however, that the second equation can easily be changed to make it ready for the substitution method:

$x - y = 11$

$x = y + 11$

DEAD ENDS

In Problem 3, we could have changed the first equation to make it ready for substitution. However, the revised first equation would produce a fractional form. If possible, avoid any calculations with fractions by changing the form of the other equation.

We will call the last equation the "revised" equation and substitute it for x into the first equation:

$2x + 3y = 27$

$2(y + 11) + 3y = 27$

We again obtained an equation with one variable. Open parentheses and collect like terms on the left:

$2y + 22 + 3y = 27$

$5y + 22 = 27$

Isolate the variable and collect like terms on the right:

$5y = 27 - 22$

$5y = 5$

Divide both sides by 5:

$y = 1$

To find the value of x, substitute 1 for y into one of the original equations. I substitute y into the second equation here:

$x - y = 11$

$x - 1 = 11$

$x = 12$

Solution set: $\{(12, 1)\}$

When using the substitution method, sometimes fractions are unavoidable, since the revision of both equations will produce fractions. Let's illustrate this with the next example.

Sample Problem 4

$$\begin{cases} 3x + 7y = 10 \\ 2x + 5y = -13 \end{cases}$$

Looking ahead, we can predict that both revised equations will produce a fractional form. All variable terms have coefficients in front of them, and in order to isolate them by themselves, we would need to divide by these coefficients. Therefore, let's merely pick the second equation and solve it for x:

Isolate the term with the variable x:

$2x + 5y = -13$

$2x = -5y - 13$

Divide both sides by 2:

$$\frac{2x}{2} = \frac{-5y - 13}{2}$$

$$x = \frac{-5y - 13}{2}$$

Now we can substitute the right side of the last expression for x into the first equation:

$3x + 7y = 10$

$$3\left(\frac{-5y - 13}{2}\right) + 7y = 10$$

$$\frac{-15y - 39}{2} + 7y = 10$$

Multiply both sides by the least common denominator (LCD), which is 2, to eliminate fractions:

$$\frac{2(-15y - 39)}{2} + 14y = 20$$

$-15y - 39 + 14y = 20$

Isolate the variable and collect like terms on both sides:

$-15y + 14y = 39 + 20$

$-y = 59$

Divide both sides by -1:

$y = -59$

To find the value of x, substitute -59 for y into one of the original equations. I use the second equation here:

$2x + 5y = -13$

$2x + 5(-59) = -13$

$2x - 295 = -13$

$2x = 295 - 13 = 282$

Divide both sides by 2:

$x = 141$

Solution set: $\{(141, -59)\}$

The last solution was neither easy nor short. Can we use some other method to shorten the solution? Yes, we can. Let's get acquainted with the second method of solving systems of linear equations.

The Addition Method

The *extended addition property of equality* is the basis for another method of solving systems of linear equations, called the addition method.

DEFINITION

The **extended addition property of equality** states that two equations can be added and the resulting equation will be equivalent to the original ones:

For all real $a, b, c,$ and $d,$ if $a = b,$ and $c = d,$ then $a + c = b + d.$

The essence of the addition method of solving systems of linear equations is to add the two equations to eliminate one of the variables. The next few examples illustrate this point.

Sample Problem 5

Solve the system:

$$\begin{cases} x + y = 14 \\ x - y = 8 \end{cases}$$
$$\overline{ 2x = 22}$$

The last line is the result of adding the two original equations. Notice that the terms with the variable y cancel each other.

WORTHY TO KNOW

The addition method of solving systems of linear equations is sometimes called the elimination method.

Divide both sides of the new equivalent equation by 2:

$x = 11$

To find y, substitute the value of 11 for x into one of the original equations:

$x + y = 14$

$11 + y = 14$

$y = -11 + 14$

$y = 3$

Solution set: $\{(11, 3)\}$

The next example illustrates that sometimes it is necessary to change the form of one of the original equations before using the addition method.

Sample Problem 6

$$\begin{cases} x = y - 22 \\ x + y = -4 \end{cases}$$

Before using the addition method, change the form of the first equation to make the system suitable for the addition method:

$$\begin{cases} y = x - 22 \\ x + y = -4 \end{cases}$$

	Subtract x from both sides	$-x + y = -22$
	Leave unchanged	$x + y = -4$
	Add two equations	$2y = -26$

Divide both sides by 2:

$y = -13$

A WORD OF ADVICE

After finding the value of one of the variables, substitute this value into one of the original equations to find the value of the other variable. Pick the equation that is easiest to solve.

Substitute –13 for y in one of the original equations to find x:

$x + y = -4$

$x + (-13) = -4$

$x - 13 = -4$

$x = 13 - 4$

$x = 9$

Solution set: $\{(9, -13)\}$

Often the multiplication property of equality needs to be applied first so that the addition of the system's equation will eliminate one of the variables.

Sample Problem 7

$$\begin{cases} 2x + 5y = 23 \\ 3x - y = 9 \end{cases}$$

Notice that adding the equations as they are would not eliminate either variable. However, we should observe that applying the multiplication property and multiplying the second equation by 5, and then adding the first unchanged equation to the newly formed but equivalent second equation will eliminate the variable y.

$\begin{cases} 2x + 5y = 23 \\ 3x - y = 9 \end{cases}$	Leave unchanged	$2x + 5y = 23$
	Multiply both sides by 5	$15x - 5y = 45$
	Add two equations	$17x \quad\ = 68$
		$x \quad\ = 4$

Substitute 4 for x in the second equation (since it is the easiest):

$3x - y = 9$

$3(4) - y = 9$

$12 - y = 9$

$-y = -12 + 9 = -3$

$y = 3$

Solution set: {(4, 3)}

Frequently, the multiplication property needs to be applied to both equations before adding them. This is usually the case when both equations have different integer coefficients in front of all terms with variables. Our last example is of this type.

Sample Problem 8

$$\begin{cases} 2x + 3y = 18 \\ 3x + 2y = 17 \end{cases}$$

First, let's decide that we will eliminate the x variable. To do this, we need to multiply both equations by some numbers so that the first terms of the changed but equivalent equations will be opposites and will cancel each other.

A WORD OF ADVICE

Instead of eliminating the variable x in Problem 8, we could have eliminated the variable y. Either approach works; pick the one that involves easier computations.

The good guess here is to multiply both sides of the first equation by 3 so that the first term of the changed but equivalent equation will be $6x$. Then multiply both sides of the second equation by -2 so that the first term of the changed but equivalent equation is $-6x$. The two terms are opposites and therefore cancel each other.

$$\begin{cases} 2x + 3y = 18 \\ 3x + 2y = 17 \end{cases}$$
$\quad$ Multiply both sides by 3 $\qquad\quad 6x + 9y = 54$
$\quad$ Multiply both sides by -2 $\qquad -6x - 4y = -34$

$\qquad\qquad$ Add two equations $\qquad\qquad\quad 5y = 20$

$\qquad\qquad\qquad\qquad\qquad\qquad\qquad\qquad\qquad y = 4$

To find x, substitute 4 for y in the first original equation:

$2x + 3y = 18$

$2x + 3(4) = 18$

$2x + 12 = 18$

$2x = 18 - 12 = 6$

$x = 3$

Solution set: $\{(3, 4)\}$

Both the addition and the substitution method can be used to obtain solutions for systems of linear equations with two variables. Which you choose depends on the original form of both equations and your effort to avoid fractions as much as possible.

The Least You Need to Know

- The key to the substitution method is eliminating a variable using substitution.
- Adding equations to eliminate a variable is the key to the addition method.
- When using the addition method, the multiplication property sometimes needs to be applied to both equations.
- Choosing the right method makes finding the solution easier.

Ratio, Proportions, and Percentage Problems

In everyday life, much of our use of mathematics involves comparisons (How much more does this pair of shoes cost than the other one?) and change (Have the house prices in my community changed during the last 5 years?). It is convenient to compare amounts and express changes using mathematical tools called ratios, proportions, and percents.

We constantly deal with these mathematical concepts. For example, percents are used in the Olympics to determine a team's winning percentage (number of wins/number of games played).

The concept of ratio is used in construction of cars: the so-called gear ratio defines the relationships between the numbers of teeth on two gears. If, for example, the smaller gear has 13 teeth, while the second, larger gear has 21 teeth, the gear ratio is 21 to 13.

Suppose some business opens a new office, but instead of 48 employees (as at the original office) this one will have only 36 employees. If they know how much paper, staples, and other office supplies they need for a year in the original office, they can calculate how much to order for the new office using the concept of proportion.

In this part, we deal with the notions of ratio, proportion, and percent and tackle problems related to these mathematical tools.

Getting Rational with Ratio Problems

In This Chapter

- The use of ratios in everyday mathematics
- How ratios differ from fractions
- Changing one ratio into another
- Solving problems with mixtures

The word *ratio* originates from the Latin word meaning "reason," from which *rational* is also derived. The use of ratios dates back to the ancient world, where they played an important role in Greek and Babylonian mathematics. For instance, Greeks used ratios in astronomy to predict eclipses and in architecture to build their temples.

The ratio is a very useful notion in everyday mathematics. A ratio gives the relative sizes of two sets but not the actual numbers of objects in those sets. For example, the fact that the ratio of green marbles to red marbles in a box is 2 to 3 tells us that for every 2 green marbles there are 3 red marbles; however, it does not tell us the number of green or red marbles.

The order of terms in a ratio is important. Notice that in the expression "the ratio of green marbles to red marbles," "green marbles" came first. This order is important: whichever word comes first, its number must come first as well. If the expression had been "the ratio of red marbles to green marbles," then the numbers would have been "3 to 2."

People constantly use ratios in everyday life. For example, to produce good concrete, the ratio of water to cement must be 1:4. This means that the weight of cement in the mixture must be 4 times the weight of the water. In biology and medicine, ratios are used to produce simple dilutions. For example, a 1:4 dilution means combining 1 unit volume of the material to be diluted to 4 unit volumes of the solvent medium.

> **DEFINITION**
>
> For any two positive numbers *a* and *b,* the **ratio** of *a* to *b* is a fraction, $\frac{a}{b}$. This ratio is also written as *a:b.*

Ways to Write Ratios

We can write ratios in several different ways. Consider an example. A car dealer orders SUVs and sedans based on the colors his customers like to buy.

Color	SUVs	Sedans
Black	6	10
White	6	12
Blue	2	6
Total	14	28

Here are some ways ratios show a comparison between the cars.

Ways to Write	Compare Black Sedans to All Sedans	Compare Blue Sedans to Blue SUVs	Compare All Sedans to All SUVs
To	10 to 28	6 to 2	28 to 14
Colon	10:28	6:2	28:14
Fraction	$\frac{10}{28}$ or $\frac{5}{14}$	$\frac{6}{2}$ or $\frac{3}{1}$	$\frac{28}{14}$ or $\frac{2}{1}$

Since ratios can be written in fraction form, they can also be simplified the same way we simplify fractions. For example, the ratio 5:25 written as a fraction can be simplified as: $\frac{5}{25} = \frac{1}{5}$. We can state that two ratios, 5:25 and 1:5, are equal ratios, since they can be written as equivalent fractions. The value of equal ratios is the same.

At the same time, ratios differ from fractions: a fraction's denominator always tells how a whole is divided into equal parts, while, with ratios, the denominator may tell one of three different things. Let's look at the table and discuss this.

Andrew's Class	Boys 22	Girls 18	Total 40
Lily's Class	Boys 23	Girls 19	Total 42

If we consider the ratio of girls in Andrew's class to all students in his class, then in the ratio $\frac{18}{40}$, the denominator describes the whole class. If we consider the ratio of girls in Andrew's class to girls in Lily's class, then in the ratio $\frac{18}{19}$, the denominator tells the number of parts in a different whole (the girls in Lily's class). Finally, if we consider the ratio of girls in Andrew's class to boys in Andrew's class, then in the ratio $\frac{18}{22}$, the denominator describes a different part of the class.

DEAD ENDS

A fraction compares things that have the same units. A ratio may compare things with the same units, like cups of orange juice to cups of water, and also compares things that don't have the same units, like 350 miles traveled to 1 tank of gas. Ratios are not like fractions when it comes to computing, so don't add or subtract ratios.

Two-Term Ratio Problems

The most common ratio problems involve a comparison between two quantities. These ratios are called two-term ratios. We need to be aware of three main considerations when doing ratio problems. First, if necessary, we have to change the quantities to the same units; then we must reduce the ratio to its simplest form. For example, what is the ratio of 6 minutes to 8 hours?

First, change the hours to minutes:

8 hours = 8 · 60 = 480 minutes

Write the ratio as a fraction and simplify:

$$\frac{6}{480} = \frac{1}{80}$$

We found that the ratio of 6 minutes to 8 hours is 1:80.

The second important point to remember while solving word problems is to write the items in the ratio in fraction form.

Finally, we must always make sure that we have the same items in the numerator and denominator. For example, if the ratio of Olga's classical CDs to her rock CDs is 14 to 25, the right setup is this:

$$\frac{\text{classical}}{\text{rock}} = \frac{14}{25}$$

This setup is wrong:

$$\frac{\text{classical}}{\text{rock}} = \frac{25}{14}$$

Now let's solve several two-term ratio problems.

Sample Problem 1

In a bag of blue and yellow candies, the ratio of blue candies to yellow candies is 3:5. If the bag contains 60 yellow candies, how many blue candies are there?

Step 1: Let the number of blue candies be x. Next, write the items in the ratio as a fraction:

$$\frac{\text{blue}}{\text{yellow}} = \frac{3}{5} = \frac{x}{60}$$

Step 2: Solve the equation by cross-multiplication:

$3 \cdot 60 = 5x$

$180 = 5x$

Divide both sides by 5:

$x = 36$

Solution: There are 36 blue candies in the bag.

Sample Problem 2

A room is 16 feet, 8 inches long, and the ratio of the length to the width is 4 to 5. What is the width of the room?

Step 1: Since the length is given in both feet and inches, let's convert it to inches using the fact that 1 foot equals 12 inches. To find how many inches are in 16 feet, we multiply 16 feet by 12 inches:

16 feet, 8 inches = $(16 \cdot 12) + 8 = 192 + 8 = 200$

We found that the length is 200 inches.

Step 2: Let x represent the width. We can now set up the equation:

$$\frac{\text{width}}{\text{length}} = \frac{4}{5} = \frac{x}{200}$$

Step 3: Solve the equation by cross-multiplication:

$4 \cdot 200 = 5x$

$800 = 5x$

Divide both sides by 5:

$x = 160$ inches

We found that the width is 160 inches.

Step 4: Let's now convert inches to feet so that the units for the width are consistent with the units for the length. Since 1 foot is 12 inches, we divide 160 inches by 12 to find out how many feet are in 160 inches:

160:12 = 13, with the remainder of 4 inches. The width is 13 feet, 4 inches.

Solution: The room width is 13 feet, 4 inches.

Sample Problem 3

A school has 300 students. If the ratio of boys to girls is 31 to 44, how many more girls are there in the school?

Step 1: Since we don't know the exact number of either boys or girls, we can't set up the equation right away. We assign variables first. Let x be the number of girls; then the number of boys will be $300 - x$, since there are 300 students overall.

Step 2: Now we can set up the equation by writing the number of students in the ratio as a fraction:

$$\frac{\text{boys}}{\text{girls}} = \frac{31}{44} = \frac{300 - x}{x}$$

Solve the equation by cross-multiplication:

$31x = 44(300 - x)$

Distribute the right side:

$31x = 13{,}200 - 44x$

Isolate the variable and collect like terms:

$31x + 44x = 13{,}200$

$75x = 13{,}200$

Divide both sides by 75:

$x = 176$

We found that there are 176 girls.

Step 3: In order to find the number of boys, subtract the number of girls from the number of all students:

$300 - 176 = 124$

We found that there are 124 boys in the school.

Step 4: By subtracting the number of boys from the number of girls, we can find out how many more girls there are in the school:

$176 - 124 = 52$

Solution: There are 52 more girls than boys in the school.

Sometimes the ratio of the objects is not given. We need to use the actual number of objects to find the ratio. For example, find the ratio of boys to girls in a class if there are 60 boys and 75 girls.

Let's write the ratio in fraction form and then simplify by the common factor of 15:

$$\frac{\text{boys}}{\text{girls}} = \frac{60}{75} = \frac{4}{5}$$

So the ratio of boys to girls in class is 4:5.

WORTHY TO KNOW

One of the most famous buildings in the world, the Parthenon in Greece, was designed using the golden ratio. Artists of all times used it because it creates shapes that are pleasing to the eye. The Greek letter ϕ (pronounced as *phi,* like in the word "fly") is used to represent the golden ratio.

Sample Problem 4

Bob has 60 marbles, 36 of which are blue and 24 of which are yellow. Dora has 40 marbles, all of them either blue or yellow. If the ratio of the blue marbles to the yellow marbles is the same for both Bob and Dora, then Bob has how many more yellow marbles than Dora?

Step 1: We know that Dora has 40 marbles, all of them either blue or yellow, but we don't know the exact numbers. We need to assign variables: Let x be yellow marbles for Dora; then the number of blue marbles for her will be $40 - x$, since she has 40 marbles overall.

Step 2: Since the ratio of blue marbles to yellow marbles is the same for both, let's obtain the needed ratio from Bob. He has 60 marbles, 36 of which are blue and 24 of which are yellow. We write the ratio 36:24 as a fraction and simplify:

$$\frac{\text{blue}}{\text{yellow}} = \frac{36}{24} = \frac{3}{2}$$

We found that the ratio of blue marbles to yellow marbles for Bob is 3 to 2.

Step 3: We use the same ratio for Dora to set up the equation:

$$\frac{\text{blue}}{\text{yellow}} = \frac{3}{2} = \frac{40 - x}{x}$$

Solve the equation by cross-multiplying:

$3x = 2(40 - x)$

Distribute the right side:

$3x = 80 - 2x$

Isolate the variable and collect like terms:

$3x + 2x = 80$

$5x = 80$

Divide both sides by 5:

$x = 16$

We found that Dora has 16 yellow marbles.

Step 4: To find how many more yellow marbles Bob has than Dora, we subtract the number of yellow marbles that Dora has from the number of yellow marbles that Bob has:

$36 - 16 = 20$

Solution: Bob has 20 more yellow marbles than Dora.

Let's do the last problem in this section.

Sample Problem 5

At a small college, the ratio of men to women is 9:4. If there are presently 720 women, how many additional women would it take to reduce the ratio of men to women to 2:1?

Step 1: To answer the problem's question, we need to know the present number of men at the college. Let this number be x. We use the first ratio 9:4 and the present number of women to find the present number of men:

$$\frac{\text{men}}{\text{women}} = \frac{9}{4} = \frac{x}{720}$$

Solve the equation by cross-multiplying:

$9(720) = 4x$

$6{,}480 = 4x$

Divide both sides by 4:

$x = 1{,}620$

We found that there are presently 1,620 men at the college.

Step 2: To reduce the ratio to 2:1, the college needs to take in more women. Let that number of additional women be y. We use the variable y since we already used the variable x to represent the amount of men. The present number of women is 720, so the new amount of women will be $720 + y$.

Step 3: Now, we know the present amount of men (this should not be changed) and we know the number of women $(720 + y)$ it will take to change the ratio to the desirable ratio 2:1, so we can set up our equation:

$$\frac{\text{men}}{\text{women}} = \frac{2}{1} = \frac{1{,}620}{720 + y}$$

Step 4: Solve the equation by cross-multiplication:

$2(720 + y) = 1{,}620$

Distribute and isolate the variable:

$1{,}440 + 2y = 1{,}620$

$2y = 1{,}620 - 1{,}440$

Collect like terms:

$2y = 180$

Divide both sides by 2:

$y = 90$

Solution: It would take 90 additional women to reduce the ratio of men to women to 2:1.

A WORD OF ADVICE

In a class of 30 students, the ratio 12 to 18 refers to the absolute numbers of girls and boys, respectively. The simplified ratio 2 to 3 tells us only that, for every 2 girls, there are 3 boys. It also tells that, in any representative set of 5 students (2 + 3 = 5) from this class, 2 will be girls and 3 will be boys. In other words, the girls comprise $\frac{2}{5}$ of the class. Use this kind of reasoning when solving ratio problems.

Three-Term Ratio Problems

A three-term ratio can be used to compare three quantities. For example, we can use the dimensions of a box to make a three-term ratio. We can write the measurements of the width, length, and height in a ratio (15:30:10) and then reduce to simplest terms by dividing all terms by 5 as 3:6:2. In three-term ratios, order is also very important. In the preceding example, the ratio of the width to the height is 3:2, and the ratio of the height to the length is 2:6.

Let's do several problems and see how it works.

Sample Problem 6

A special cereal mixture contains wheat, corn, and rice in the ratio of 2:3:5. If a bag of the mixture contains 6 pounds of wheat, what is the weight of the whole cereal mixture bag?

Step 1: To find the weight of the whole bag, we need to find the weights of all three components. We already know the weight of wheat, but the weights of corn and rice are unknown. We need to assign variables. Let the amount of corn be x and the amount of rice be y.

Step 2: The ratio of wheat to corn is 2:3. There are 6 pounds of wheat and x pounds of corn. We write the items in the ratio as a fraction:

$$\frac{\text{wheat}}{\text{corn}} = \frac{2}{3} = \frac{6}{x}$$

Solve the equation by cross-multiplication:

$2x = 3 \cdot 6$

$2x = 18$

Divide both sides by 2:

$x = 9$

We found that there are 9 pounds of corn in the mixture.

Step 3: Next, we find the amount of rice. The ratio of wheat to rice is 2:5. There are 6 pounds of wheat and y pounds of rice. We write the items in the ratio as a fraction:

$$\frac{\text{wheat}}{\text{rice}} = \frac{2}{5} = \frac{6}{y}$$

Solve by cross-multiplication:

$2y = 5 \cdot 6$

$2y = 30$

Divide both sides by 2:

$y = 15$

There are 15 pounds of rice in the mixture.

Step 4: To find the weight of the cereal bag, we need to add the weights of all three components:

$6 + 9 + 15 = 30$

Solution: The weight of the cereal bag is 30 pounds.

Sample Problem 7

A store at the mall sells skirts in only three colors: black, blue, and brown. The colors are in the ratio of 3 to 4 to 5. If the store has 40 blue skirts, how many more brown skirts than black skirts does it have?

Step 1: To answer the problem's question, we need to know the number of black and brown skirts. First we assign variables. Let x be the number of black skirts, and let y be the number of brown skirts; we already know the number of blue skirts (40).

Step 2: Next, we find the number of black skirts. The ratio of black skirts to blue skirts is 3 to 4. The number of blue skirts is 40 and the number of black skirts is x. We can write the items in the ratio as a fraction:

$$\frac{\text{black}}{\text{blue}} = \frac{3}{4} = \frac{x}{40}$$

Solve the equation by cross-multiplication:

$4x = 3 \cdot 40$

$4x = 120$

Divide both sides by 4:

$x = 30$

There are 30 black skirts in the store.

WORTHY TO KNOW

You've heard the words *sine, cosine,* and *tangent.* These terms indicate special relationships between angle measures and side lengths in right triangles. These relationships are called trigonometric ratios.

Step 3: Next, we find the number of brown skirts. The ratio of the blue skirts to the brown ones is 4 to 5, and there are 40 blue skirts and y brown skirts. We can now write the items in the ratio as a fraction:

$$\frac{\text{blue}}{\text{brown}} = \frac{4}{5} = \frac{40}{y}$$

Solve the equation by cross-multiplication:

$4y = 5 \cdot 40$

$4y = 200$

Divide both sides by 4:

$y = 50$

There are 50 brown skirts at the store.

Step 4: We found that there are 30 black skirts and 50 brown skirts at the store. By subtracting the number of black skirts from the number of brown skirts, we can find how many more brown skirts there are:

$50 - 30 = 20$

Solution: There are 20 more brown skirts than black skirts at the store.

Practice Problems

Problem 1: In a rectangle, the ratio of the width to the length is 3:4. What is the length of the rectangle if the width is 3 feet, 3 inches?

Problem 2: In a healthy person, the ratio of red blood cells to other blood cells should be around 1 to 5,000. A man learned from his doctor that his count is 300 red blood cells out of 240,000 blood cells. Is his count low, high, or normal?

Problem 3: The ratio of dentists who recommended a new toothpaste to the number of all dentists interviewed is 5 to 8. If 528 dentists were interviewed, how many recommended the new toothpaste?

Problem 4: Lora has 75 marbles, 45 of which are red and 30 of which are green. Anna has 40 marbles, all of them either red or green. If the ratio of the red marbles to the green marbles is the same for both Lora and Anna, then Lora has how many more green marbles than Anna?

Problem 5: In a college, the ratio of men to women is 7 to 5. If there are 840 more men than women, what is the total enrollment?

Problem 6: At the math department, the ratio of males to females among the 500 incoming freshmen is 7 to 3. How many additional women should be accepted into the department to achieve the desired ratio of 2:1?

Problem 7: The number of pigs, goats, and hens on a farm is in the ratio 2:3:10, respectively. If there are 120 hens, how many more goats than pigs are there?

The Least You Need to Know

- Ratios can be written in three different ways (written as 1 to 2; with a colon, 1:2; and as a fraction, $\frac{1}{2}$).
- Equivalent ratios have the same value.
- Write the items in the ratio as a fraction to set up the equation.
- The most useful method to solve an equation for ratio problems is cross-multiplication.

Dealing With Proportions

In This Chapter

- Writing proportions
- Finding common elements for numerators and denominators
- Finding distances on maps
- Finding the height of an object without measuring it directly

We deal with proportions every day. When we cook for more people than a recipe suggests, we need to increase the amount of each ingredient proportionally; otherwise, the dish will taste different. When we enlarge a wallet-size photo to a bigger one to hang up on a wall, we recognize ourselves immediately, since the features of our face are still in the same proportion to one another.

In this chapter, we'll talk about word problems that involve *proportions*, discuss three methods of solving proportions, and learn how to set up directly proportional and inversely proportional problems.

What Are Proportions?

We can write proportions in different ways. Consider the following example. The height of the famous Eiffel Tower is a little more than 300 m, and its base length is around 100 m. So the ratio of the height to the length is 3:1. If a souvenir maker wants to make an Eiffel Tower that is 30 cm high, in order for the souvenir to reflect the actual dimensions, the length of the souvenir base must be 10 cm.

> **DEFINITION**
>
> A **proportion** is an equation showing that two ratios are equal.

To write the correct proportions about this relationship, some common element must connect the numerators. Another common element connects the denominators. For example, both numerators can relate to the scale dimensions, while both denominators can relate to the actual dimensions:

$$\frac{\text{height of souvenir}}{\text{actual height}} = \frac{\text{base length of souvenir}}{\text{actual base length}}$$

Or, both numerators could relate to one dimension, such as height, while both denominators could relate to the base length:

$$\frac{\text{height of souvenir}}{\text{base length souvenir}} = \frac{\text{actual height}}{\text{actual base length}}$$

We can write these proportions using a colon:

30:300 = 10:100

30:10 = 300:100

We also can write them in fraction form:

$$\frac{30}{300} = \frac{10}{100}$$

$$\frac{30}{10} = \frac{300}{100}$$

In the proportion $a{:}b = c{:}d$, a and d are called the extreme terms, and b and c are called the mean terms. The product of the extreme terms equals the product of the mean terms: $ad = bc$.

Example: Is 3:9 = 5:15 a true proportion? The product of the extremes is $3 \cdot 15 = 45$; the product of the means is $5 \cdot 9 = 45$. So 3:9 = 5:15 is a true proportion.

Directly Proportional Problems

When two quantities are directly proportional, a change in one quantity causes a predictable change in the other. This means that both quantities either increase or decrease by the same factor.

Example 1: Your hourly wage is $10. If you increase your hours by the factor of 5, your total earnings will increase by the same factor and be $50. So hours and total earnings are directly proportional.

Example 2: Prices for produce are usually listed by the pound. A pound of tomatoes may cost $2.40. But you can buy only one big tomato, which weighs $\frac{1}{3}$ of a pound.

You decreased the weight by the factor of 3. As the weight of the produce decreases, the price you pay decreases by the same factor. You will pay 80¢ for one tomato. So weight and price are also directly proportional.

Methods of Solving Proportions

We can use three methods for solving proportions. Let's illustrate these with an example:

Solve the proportion:

$$\frac{3}{1,200} = \frac{x}{3,600}$$

Method 1: We can solve a proportion by finding equivalent fractions:

$$\frac{3}{1,200} = \frac{3 \cdot 3}{1,200 \cdot 3} = \frac{9}{3,600} = \frac{x}{3,600}$$

Two fractions have the same denominator, so in order for them to be equal, the numerators must be the same as well.

x = 9

Method 2: We can also use what we know about the product of extremes and means. First, we write the proportion using a colon:

3:1,200 = *x*:3,600

Now we calculate the products:

3 · 3,600 (product of extremes) = 1,200*x* (product of means)

10,800 = 1,200*x*

Divide both sides by 1,200:

x = 9

Method 3: We can finally solve a proportion by cross-multiplication:

$$\frac{3}{1,200} = \frac{x}{3,600}$$

$1,200x = 3 \cdot 3,600$

$1,200x = 10,800$

$x = 9$

No matter what method we choose, the answer will be the same.

Note that the most common method of solving proportions is cross-multiplication.

WORTHY TO KNOW

The relationship between the extremes and the means is known as the Rule of Three, because if we know any three terms in a proportion, we can find the fourth.

How to Set Up a Proportion

Math problems involving proportions have three known values and an unknown value. The unknown is what you are asked to find.

Directly proportional problems are usually of this form: If a_1, then b_1. If a_1 changes to a_2, then the new value of b_1 will be b_2.

Consider the following example. If two pens cost \$2.50, how many pens can we buy with \$20? Let's assign variables. Since we know both amounts of money, let the original amount of \$2.50 be a_1. Then the new amount of money (\$20) will be a_2. The original two pens will be b_1, and the new number of pens will be x, since we need to find it.

At the beginning of the chapter, we discussed that, in order to write the correct proportion for directly proportional problems, some common element must connect the numerators. Another common element connects denominators. For example, both numerators could relate to the money, while both denominators could relate to the number of pens:

$$\frac{\text{original amount of money}}{\text{original number of pens}} = \frac{\text{new amount of money}}{\text{new number of pens}}$$

Then the setup for the proportion is this:

$$\frac{a_1}{b_1} = \frac{a_2}{x}$$

Plug in the known values:

$$\frac{2.5}{2} = \frac{20}{x}$$

Cross-multiply:

$2.5x = 2 \cdot 20$

$2.5x = 40$

Divide both sides by 2.5:

$x = 16$ pens

Another option is that both numerators could relate to the original values, while both denominators relate to the new values:

$$\frac{\text{original amount of money}}{\text{new amount of money}} = \frac{\text{original number of pens}}{\text{new number of pens}}$$

Then the setup for the proportion is this:

$$\frac{a_1}{a_2} = \frac{b_1}{x}$$

Plug in the known values:

$$\frac{2.5}{20} = \frac{2}{x}$$

Cross-multiply:

$2.5x = 20 \cdot 2$

$2.5x = 40$

Divide both sides by 2:

$x = 16$ pens

As we can see, both setups provided us, as we expected, with the same answer. We will buy 16 pens.

> **A WORD OF ADVICE**
>
> In directly proportional problems, the original or initial values for both items or objects must be aligned either in the same row:
>
> $$\frac{a_1}{a_2} = \frac{b_1}{b_2}$$
>
> or in the same column:
>
> $$\frac{a_1}{b_1} = \frac{a_2}{b_2}$$

Now we are fully equipped to do several directly proportional problems.

Sample Problem 1

Sara finds that she can read 18 pages in a book in 20 minutes. At this rate, how long will it take her to finish a book that is 468 pages long?

Step 1: Let's first assign variables. Since both numbers of pages are given, the original 18 pages will be a_1, and 468 pages will be a_2. The original time of 20 minutes will be b_1, and the time it will take Sara to read 468 pages will be x.

Step 2: The common element for the numerators will be pages, and the common element for the denominators will be minutes. We are now ready to write the proportion:

$$\frac{a_1}{b_1} = \frac{a_2}{x}$$

Plug in the known values from Step 1:

$$\frac{18}{20} = \frac{468}{x}$$

Step 3: Solve the proportion by cross-multiplication:

$18x = 20 \cdot 468$

$18x = 9,360$

Divide both sides by 18:

$x = 520$

We found that it will take Sara 520 minutes to finish the book.

Step 4: Since 520 minutes is definitely more than an hour, we need to convert minutes into hours. Since 1 hour is 60 minutes, we need to divide 520 minutes by 60 minutes to find how many hours are in 520 minutes:

520:60 = 8 hours, with the remainder of 40

Solution: It will take Sara 8 hours and 40 minutes to finish the book.

Sample Problem 2

If 1.5 cups of flour are required to make 30 cookies, how many cups of flour are required for 100 cookies?

Step 1: Let's assign variables first. The initial amount of flour, which is 1.5 cups, will be a_1, and the initial amount of 30 cookies that we can make with it will be b_1. The increased amount of flour that we don't know will be x, and the new amount of cookies, which is 100, will be b_2.

Step 2: Now we can set up the proportion. The common element for the numerators will be the amount of flour, and the common element for the denominators will be the number of cookies we can make with this amount of flour:

$$\frac{\text{initial amount of flour}}{\text{initial number of cookies}} = \frac{\text{changed amount of flour}}{\text{changed number of cookies}}$$

Plug in the variables from Step 1:

$$\frac{a_1}{b_1} = \frac{x}{b_2}$$

Step 3: Now we can plug in the actual values into the proportion:

$$\frac{1.5}{30} = \frac{x}{100}$$

Solve the proportion by cross-multiplication:

$30x = 1.5 \cdot 100$

$30x = 150$

Divide both sides by 30:

$x = 5$

Solution: We need 5 cups of flour to make 100 cookies.

Many proportional problems deal with the *scale drawing*. Let's do one of these problems.

DEFINITION

A **scale drawing** is an accurate picture of something but is of a different size. The ratio of the picture's size to the actual measure is called scale.

Sample Problem 3

In a scale drawing of the Simpsons' apartment, the apartment length is 3 inches. If the scale of the drawing is 1 inch = 12 feet, find the actual length of the apartment.

Step 1: Let's first assign the variables. Scale measure, which is 1 inch, will be a_1, and the actual measure, which is 12 feet, will be a_2. Scale length of the apartment, which is 3 inches, will be b_1, and the unknown actual length will be x.

Step 2: The common element for the numerators will be scale items, and the common element for the denominators will be actual items. We are ready to set up a proportion:

$$\frac{\text{scale measure}}{\text{actual measure}} = \frac{\text{scale length}}{\text{actual length}}$$

$$\frac{a_1}{a_2} = \frac{b_1}{x}$$

Substitute the known values:

$$\frac{1}{12} = \frac{3}{x}$$

Step 3: Solve by cross-multiplication:

$1x = 12 \cdot 3$

$x = 36$

Solution: The actual length of the apartment is 36 feet.

Every map is also a scale drawing with an appropriate scale in miles or kilometers. Let's do one of these problems.

Sample Problem 4

On a map, $\frac{1}{3}$ of an inch represents 36 miles. If two cities are $2\frac{1}{2}$ inches apart on the map, what is the actual distance between them?

Step 1: Since $\frac{1}{3}$ inch represents 36 miles, the scale of the map is $\frac{1}{3}$ in = 36 miles.

Step 2: Let's first assign the variables. Scale measure, which is $\frac{1}{3}$ inch, will be a_1, and the actual measure, which is 36 miles, will be a_2. Scale distance between the two cities, which is $2\frac{1}{2}$ inches, will be b_1, and the unknown actual distance will be x.

Step 3: For this problem, we choose common elements differently from the previous problem. Let's agree that the common element for the numerators will be measures, and the common element for the denominators will be distances. Then the setup of the proportion is this:

$$\frac{\text{scale measure}}{\text{scale distance}} = \frac{\text{actual measure}}{\text{actual distance}}$$

$$\frac{a_1}{b_1} = \frac{a_2}{x}$$

In this proportion, the original values are aligned in the column.

$$\frac{\frac{1}{3}}{2\frac{1}{2}} = \frac{36}{x}$$

Step 4: This proportion contains fractions, but the rules are still the same. We solve the proportion by cross-multiplication:

$$\frac{1}{3}x = 2\frac{1}{2} \cdot 36$$

Convert the mixed number into an improper fraction on the right side:

$$\frac{x}{3} = \frac{5}{2} \cdot 36$$

$$\frac{x}{3} = \frac{5 \cdot 36}{2}$$

$$\frac{x}{3} = \frac{180}{2}$$

We simplified the original proportion and got rid of the fractions. Now we can cross-multiply one more time to find x:

$2x = 3 \cdot 180$

$2x = 540$

Divide both sides by 2:

$x = 270$

Solution: The two cities are 270 miles apart.

Proportions help to find the missing measures of *similar figures*. When an object is high enough to make the measurement of its height an inconvenient matter, we can use similar right triangles to find the height.

DEFINITION

Similar figures are figures that have the same shape but are different in size. Corresponding sides of similar figures are in proportion, and corresponding angles are congruent (having exactly the same size and shape).

Ancient scientists knew how to measure the height of the Great Pyramid in Egypt. Instead of climbing to the top and measuring its height directly, they invented a very clever and much safer method of measuring its shadow. They used a stick that they pushed into the ground so it stood straight and sturdy. Then they waited until the sun was high in the sky and the stick's shadow equaled the length of the stick. They further reasoned that, in the same manner, at this exact moment, the height of the pyramid would equal the length of its shadow, and measuring the shadow would be the same as measuring the height itself, only easier.

The described method requires patience, precision, and a lot of vacant and flat space around the measured object. Another method of measurement also employs shadows, but we don't have to wait until the length of the stick and the length of its shadow are equal. We just need to measure the length of the stick and the length of its shadow, and then immediately measure the length of the shadow cast by the tall object. The tall object and its shadow form a right-angle triangle. The stick and its shadow form a similar right-angle triangle. Thus, the height of the tall object can be calculated by solving a proportion.

Sample Problem 5

Kevin is 6 feet tall, and his shadow is 10 feet long. How high is the tree whose shadow is 90 feet long?

Step 1: When we discussed the "original or initial values" in proportion problems, we had a very broad meaning in mind. For example, in the case of this problem, what can be considered the original values? We would say the boy's and tree's heights. The reason they even have shadows is because they had their heights. But the rule stays the same: the heights must be aligned either in the same row or in the same column. The same is true for the lengths of shades.

Step 2: Now we are ready to assign variables. Let the boy's height be a_1 and the length of his shadow be b_1. Since the height of the tree is unknown, let it be x. The length of the tree shadow will be b_2.

Step 3: Draw a picture of the similar triangles formed by the boy and his shadow and by the tree and its shadow to visualize the problem's situation:

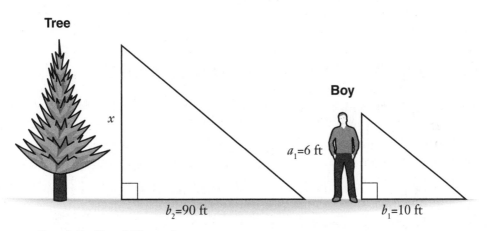

Sample Problem 5 Illustration.

Step 4: The common element for the numerators will be the heights, and the common element for the denominators will be the length of the shadows:

$$\frac{\text{boy's height}}{\text{boy's shadow length}} = \frac{\text{tree's height}}{\text{tree's shadow length}}$$

Let's substitute the variables:

$$\frac{a_1}{b_1} = \frac{x}{b_2}$$

Step 5: By plugging in the actual values for known variables, we obtain our proportion:

$$\frac{6}{10} = \frac{x}{90}$$

Solve the proportion by cross-multiplication:

$10x = 6 \cdot 90$

$10x = 540$

Divide both sides by 10:

$x = 54$

Solution: The height of the tree is 54 feet.

WORTHY TO KNOW

Thales (*THAY-leez*), a sixth-century B.C.E. Greek philosopher/mathematician/scientist, fascinated his contemporaries when he worked out a way to use a proportional triangle to tell how far a ship is from the shore.

Inversely Proportional Problems

In an inverse proportion, when one quantity increases by a certain factor, the other quantity decreases by the same factor, and vice versa.

Consider the following example. One day you run 5 miles, with a speed of 6 mph. The next day you cover the same distance on your bike, with a speed of 12 mph. The amount of time it takes to run 5 miles decreases from $\frac{5\,\text{miles}}{6\,\text{mph}} = \frac{5}{6}$ hours (50 minutes) to $\frac{5\,\text{miles}}{12\,\text{mph}} = \frac{5}{12}$ hours (25 minutes) when using the bike. You increased your speed by the factor of 2, and your time decreased by the same factor of 2.

A WORD OF ADVICE

When converting a fraction of an hour into minutes and having the denominator of the fraction as a factor of 60 (2, 3, 4, 5, 6 10, 12, 15, 20, 30), multiply both the numerator and the denominator by another factor of 60 so that the new denominator is equal to 60. For example: $\frac{7}{15}$ hours $= \frac{7 \cdot 4}{15 \cdot 4} = \frac{28}{60}$. The new numerator will show you the number of minutes; for example, $\frac{7}{15}$ hours becomes 28 minutes.

Inversely proportional problems also have three known values and an unknown value. The unknown is what you are asked to find. The problems will sound very similar to directly proportional problems. The most common example of this type is: the more people on a job, the less time it will take for the job to be completed. The setup of a proportion differs from the one we have done so far.

The right setup for the inversely proportional problems is $\dfrac{a_1}{a_2} = \dfrac{b_2}{b_1}$ or $\dfrac{a_1}{b_2} = \dfrac{a_2}{b_1}$.

Let's discuss this while doing the next problem.

Sample Problem 6

It takes 4 workers 6 hours to repair a water leak. How long will it take 7 workers to do the job if they work at the same rate?

Step 1: Let's first sort out the variables. The initial number of workers, which is 4, will be a_1; the prospective number of workers, which is 7, will be a_2. The initial number of hours, which is 6 hours, will be b_1, and the prospective number of hours that we need to find will be x.

Step 2: In the case of inverse proportion, we should not have any common element in the numerators or in the denominators. For this problem, common elements are either the initial values or the prospective values. The setup should be like this:

$$\frac{a_1}{a_2} = \frac{x}{b_1}$$

Another right setup is this:

$$\frac{a_1}{x} = \frac{a_2}{b_1}$$

So we wrote the inverse relationship.

DEAD ENDS

For inversely proportional problems, initial (original) items should never be aligned in the same row or column; otherwise, you get the directly proportional setup.

Step 3: Let's substitute the actual values in the first equation from Step 2:

$$\frac{4}{7} = \frac{x}{6}$$

Solve the equation by cross-multiplication:

$7x = 4 \cdot 6$

$7x = 24$

Divide both sides by 7:

$x = 3\dfrac{3}{7}$

Solution: It will take 7 workers $3\dfrac{3}{7}$ hours to do the job.

Sample Problem 7

If 8 water pumps can fill a tank in 9 days, how many days will it take 10 pumps to fill the same tank?

Step 1: First we assign variables. The initial number of water pumps will be a_1, and the prospective number of pumps will be a_2. The initial number of days will be b_1, and the prospective number of days will be x, since we need to find it.

Step 2: Now we can set up the proportion:

$$\frac{\text{initial number of pumps}}{\text{prospective number of pumps}} = \frac{\text{prospective number of workers}}{\text{initial number of workers}}$$

$$\frac{a_1}{a_2} = \frac{x}{b_1}$$

Step 3: Now we substitute the actual values in the equation from Step 2:

$$\frac{8}{10} = \frac{x}{9}$$

Solve by cross-multiplication:

$10x = 8 \cdot 9$

$10x = 72$

Divide both sides by 10:

$x = 7.2$

Solution: 10 water pumps will finish the job in 7.2 days.

Practice Problems

Problem 1: A worker used 3 gallons of interior paint to cover 600 ft². If he needs to paint an additional 1,800 ft², how many more gallons of paint does he need?

Problem 2: A car travels 250 miles in 5 hours. How far would it travel in 12 hours?

Problem 3: Christopher casts a shadow 5.4 m long at the same time that his little sister casts a shadow 3.3 m long. If Christopher is 1.8 m tall, how tall is his little sister?

Problem 4: On a map, 1 cm represents 65 km. If two cities are 1,950 km apart, how many centimeters apart would they be on this map?

Problem 5: If 3 tree-trimmers can trim the trees in a park in 48 work-hours, how many work-hours would it take 8 tree-trimmers to trim the trees in the park?

Problem 6: If 3 people are needed to complete a job in 5 days, how long will it take to complete the job if 10 people are used?

Problem 7: It takes 4 people 6 days to mow lawns in a community. How many days will it take 6 people to mow the same lawns?

The Least You Need to Know

- The right setup for directly proportional problems is $\frac{a_1}{b_1} = \frac{a_2}{b_2}$ or $\frac{a_1}{a_2} = \frac{b_1}{b_2}$.

- The right setup for inversely proportional problems is $\frac{a_1}{a_2} = \frac{b_2}{b_1}$ or $\frac{a_1}{b_2} = \frac{a_2}{b_1}$.

- To find the actual distance between two destinations on a map, we need to know the map's scale.

- Proportions help to find the missing measures of similar figures such as the height of an object or a person.

Percentage Solutions

In This Chapter

- Introducing the main formula for percent problems
- Converting percents into decimals, and vice versa
- Understanding what is the whole and what is a part
- Writing a word statement and its math equivalent

Percents are a useful mathematical tool that can make it easier to compare quantities. If we want to compare the number of fiction books at one library to the number of fiction books at another library, we could say that $\frac{3}{5}$ of all books at the first library are fiction and $\frac{5}{8}$ of all books at the other library are fiction. But if we use percents instead of fractions, the comparison is more obvious: 60% is less than 62.5%.

In this chapter, we introduce three types of percent problems that can be solved with one basic formula. Later in the chapter, we will learn how to calculate percent of change.

Introducing Percent

The word *percent* comes from the Latin *per centum*, meaning "out of 100." The concept of percent originated from societies' need to compute taxes, loss, profit, and interest. During the Middle Ages, as larger amounts of money were used, 100 became a common base for computations. The modern symbol for percent (%) first appeared in an anonymous Italian manuscript around the 1420s.

Calculations with percent fall into three categories:

- Given the whole and the percent, find the part.

- Given the whole and the part, find the percent.

- Given the percent and the part, find the whole.

One equation helps to solve all three categories of percent problems:

$b = p\% \cdot a$

Stating this in words:

Part of the whole equals percents of a whole.

$b = p\% \cdot a$

This algebraic percent equation has three variables: part of the whole b, percents $p\%$, and a whole a. Knowing two of those numbers, we can find the third. Remembering the following helps identify the type of percent problem we need to solve:

- The number after *of* is always the whole.

- The number before or right after *is* is always a part.

    For example, if we say 30% of __ is 50 or 50 is 30% of __, we know that the whole is missing and the part is 50. If we say __ is 15% of 60, we know that the part is missing and the whole is 60.

Note that when solving the percent equation, the percents must be always entered as a decimal.

A WORD OF ADVICE

To change a percent to a decimal:

- Drop the percent symbol, %.

- Move the decimal point two places to the left.

- Prefix zeros, if necessary.

- 1.5% = 0.015

Finding a Part of the Whole

When the whole and percents are given, we use the formula directly. When we learned to translate simple words into mathematical expressions, we discussed that *of* can stand for *times*. This frequently comes up when we solve percentage problems.

Sample Problem 1

What is 0.05% of 650?

Step 1: When solving percent problems, it is useful to first write the word statement and then the equation under it. We know the percent (0.05%) and the whole (650). The unknown is the comparative part, which we denote as x:

What is 0.05% of 650?

$x = 0.05\% \cdot 650$

Step 2: To solve the equation, we first convert the percent into decimal form and substitute it into the equation in Step 1:

$0.05\% = 0.0005$

$x = 0.0005 \cdot 650 = 0.325$

Solution: The decimal 0.325 is 0.05% of 650.

Sample Problem 2

40% of 72 is what?

Step 1: We can change the word order so that the problem sounds similar to Problem 1: What is 40% of 72? Let's say that the unknown part of 72 is x. We can now write the statement and the equation:

What is 40% of 72?

$x = 40\% \cdot 72$

Step 2: Convert the percent into decimal form and plug it into the equation in Step 1:

$40\% = 0.40$

$x = 0.40 \cdot 72 = 28.8$

Solution: 28.8 is 40% of 72.

Sample Problem 3

The Johnson family's weekly food budget is $220. 30% of it can be spent on eating out, and 5% of it can be spent on snacks at the movie theater. What amount of money does the Johnson family spend on other food?

Step 1: First we must find how much money the Johnson family spends on eating out. Let's say that this amount of money is x. We can now write the statement:

Eating out is 30% of the weekly food budget.

x = 30% · $220

Step 2: Convert the percent into decimal form and substitute into the equation in Step 1:

30% = 0.30

x = 0.30 · 220

x = 66

We found that the Johnson family spends $66 on eating out.

Step 3: Now we calculate how much money they spend on snacks. Let's denote this amount of money as y. We can state:

Snacks are 5% of the weekly food budget.

y = 5% · $220

Step 4: Convert 5% into decimal form and substitute into the equation in Step 3:

5% = 0.05

y = 0.05 · 220

y = 11

We calculated that the Johnson family spends $11 on snacks.

Step 5: To answer the problem's question, we first calculate how much money the family spends on eating out and on snacks together by adding these two amounts:

$66 + $11 = $77

The family spends $77 on eating out and snacks. To calculate how much money the Johnson family spends on other food, we subtract $77 from the weekly food budget:

$220 – $77 = $143

Solution: The Johnson family spends $143 on other food weekly.

DEAD ENDS

Remember that the number before or after *is* is always a part, and the number after *of* is always the whole. If you interchange them, the solution will be wrong.

Sample Problem 4

Selena bought a coat with 6% tax. The price before the tax was $860. What was the price of the coat with the tax?

Step 1: Tax is unknown and is calculated based on price, namely 6% of the price constitutes the tax. Let the tax be x. We can write this statement:

Tax is 6% of the coat price.

$x = 6\% \cdot \$860$

Step 2: To calculate the tax, we need to convert the percent into a decimal:

$6\% = 0.06$

Step 3: To find the tax, we need to solve the equation:

$x = 0.06 \cdot 860$

$x = 51.60$

We found that the tax was $51.60.

Step 4: To find the total price for the coat, we add the price without tax to the tax amount:

$\$860 + \$51.60 = \$911.60$

Solution: The total price for the coat was $911.60.

WORTHY TO KNOW

The word *per mille* (or *promille*) originated from the Latin word meaning "for thousand" and is a tenth of a percent:

$$\permil = \frac{1}{1000} = 0.001 = 0.1\,\%$$

It is commonly used to describe blood alcohol content, sea water salinity, birth and death rates, and property taxation in the United States.

Sample Problem 5

In a class of 2,250 students, 44% were girls. How many more boys are there in the school?

Step 1: To answer the problem's question, we first need to find the number of girls and boys in the class. The girls constitute a part (44%) of the whole class. Let's say that the number of girls in the class is x.

Step 2: We can now write this statement:

Girls are 44% of 2,250 students.

$x = 44\% \cdot 2,250$

Step 3: Now we need to convert the percent into a decimal:

$44\% = 0.44$

Step 4: Substitute the percent of girls in decimal form into the equation:

$x = 0.44 \cdot 2,250$

$x = 990$

We found that there are 990 girls in the school.

Step 5: To find the number of boys, we subtract the number of girls from the total number of all students:

$2,250 - 990 = 1,260$

We found that the number of boys in school is 1,260.

Step 6: To find how many more boys there are in the school, we subtract the number of girls from the number of boys:

$1,260 - 990 = 270$

Solution: There are 270 more boys than girls in the school.

Sample Problem 6

There were 120 people in a school football stadium. If 25% more people come and then 12% of the new total leave, how many people remain in the stadium?

Step 1: In this problem, we need to find a part of a whole two times: first, we need to calculate 25% of the original number of people (120) to find the new total; then we need to find 12% of the new total to answer the problem's question.

Step 2: To find the new total, we first need to calculate how many more people in addition to the original total (120) came to the stadium. Let this additional number of people be x. We can write this statement:

Additional people are 25% of 120 people.

$x = 25\% \cdot 120$

Step 3: To find x, we convert 25% into decimal and plug it into the equation:

$25\% = 0.25$

$x = 0.25 \cdot 120 = 30$

30 new people came to the stadium.

Step 4: To find the new total, we add the original total and the number of additional people:

$120 + 30 = 150$

We found that the new total is 150 people.

Step 5: We know that 12% of the new total left, so we need to calculate how many people constitute this 12%. Let's assign variable y to represent this amount of people. We can write this statement:

People who left are 12% of the new total.

$y = 12\% \cdot 150$

Step 6: Before we solve for y, let's convert the percent into decimal form:

$12\% = 0.12$

Plug this value into the equation:

$y = 0.12 \cdot 150$

$y = 18$

We found that 18 people left.

Step 7: To find how many people remain in the stadium, we subtract the number of people who left from the new total:

$150 - 18 = 132$

Solution: 132 people remain in the stadium.

Finding the Whole When the Percent Is Known

When we know the percent and the part of the whole, we can find the whole using the main formula in this form:

$$b = \frac{a}{p\%}$$

a is a part.

b is the whole.

$p\%$ is the percent that we need to use in its decimal form.

Sample Problem 7

45% of what is 18?

Step 1: Here we have the percent (45%) and the comparative part. The whole is unknown, and we will assign x to represent it. Our statement and its translation into a mathematical expression is this:

45% of what is 18?

$45\% \cdot x = 18$

Step 2: Convert 45% into its decimal form:

$45\% = 0.45$

Step 3: To find x, we use the percent formula in a form that we discussed at the beginning of this section:

$$x = \frac{a}{p\%} = \frac{18}{0.45} = \frac{1,800}{45} = 40$$

Solution: 18 is 45% of 40.

A WORD OF ADVICE

To divide by a decimal: Change the divisor to a whole number by moving the decimal point the necessary number of places to the right.

Move the decimal point in the dividend the same number of places to the right.

Divide as usual.

Sample Problem 8

24% of what is 72?

Step 1: We have the percent (24%) and the comparative part. The whole is unknown, and we will assign x to represent it. Our statement and its translation into mathematical expression is this:

24% of what is 72?

24% · x = 72

Step 2: Convert the percent into its decimal from and use the formula for the whole:

24% = 0.24

$$x = \frac{a}{p\%} = \frac{72}{0.24} = \frac{7,200}{24} = 300$$

Solution: 72 is 24% of 300.

Sample Problem 9

In a school, 16% of the teachers teach English. If there are 8 English teachers in the school, how many teachers are there in the school?

Step 1: We don't know how many teachers are in the school, so let's assign x to represent this number. We know that the English portion of the teachers is 16%, so we can write our statement and the equation:

English teachers are 16% of all teachers.

8 = 16% · x

Step 2: Convert 16% into its decimal form and plug it into the equation for the whole:

16% = 0.16

$$x = \frac{a}{p\%} = \frac{8}{0.16} = \frac{800}{16} = 50$$

Solution: There are 50 teachers in the school.

Sample Problem 10

Tom has a large container of orange juice. After he has used 3 quarts of juice, 85% of the juice remains. How many quarts of orange juice remained?

Step 1: To answer the problem's question, we need to calculate how many quarts of juice were in the full container. We need to find the whole, which we will represent by x.

Step 2: We know that Tom used 3 quarts of juice, but we don't know how many percent of the whole that constitutes. We know that 85% of the juice remained, so to find the percent of used juice, we subtract 85% from 100% (since the whole is always represented by 100%):

100% − 85% = 15%

We found that Tom used 15% of the juice and that it was 3 quarts.

Step 3: We know the comparative part (3 quarts) and the percent it constitutes (15%), so we can write the percent statement that will help us find the whole (full container):

3 quarts is 15% of what?

$3 = 15\% \cdot x$

Step 4: Convert the percent into its decimal form:

15% = 0.15

Now we can calculate how many quarts of juice were in the full container by using the percent formula for the whole:

$$x = \frac{3}{0.15} = \frac{300}{15} = 20$$

We found that there were 20 quarts of orange juice in the full container.

Step 5: To calculate how many quarts of juice remained, we subtract the amount of juice used (3 quarts) from the amount in the full container:

20 − 3 = 17 quarts

Solution: 17 quarts of orange juice remained.

Finding What Percent One Number Is of Another

When we know the whole and the part and need to find the percent, we use the percent formula in this form:

$$p\% = \frac{a}{b}$$

a is the part.

b is the whole.

$p\%$ is the unknown percent.

Usually, the problem asks the question that starts with "what percent," so it is easy to identify what type of percent problem this particular one is.

Sample Problem 11

What percent of 50 is 30?

Step 1: Since the word *of* stands in front of number 50, 50 is the original whole and 30 is a comparative number. The unknown is the rate or percentage. We will assign x to represent it.

Step 2: Next, we rephrase the problem so it is easier to write the math expression:

30 is what percent of 50?

$30 = x \cdot 50$

Step 3: We use the appropriate formula to find the percent:

$$x = \frac{a}{b} = \frac{\text{part}}{\text{whole}} = \frac{30}{50} = 0.6$$

Step 4: We found the answer in decimal from. We need to convert it into percent:

$0.6 = 60\%$

Solution: 30 is 60% of 50.

A WORD OF ADVICE

To change a decimal to a percent, move the decimal point two places to the right and write the percent symbol.

$0.024 = 2.4\%$

Sample Problem 12

A serving of cake contains 480 calories. 288 calories come from fat. What percent of the total calories come from fat?

Step 1: We know that the total number of calories is 480 calories, and part of it, which is 288 calories, comes from fat. We need to know what percent this fat constitutes. Let's denote this percent as x.

Step 2: We can write the statement and the equation:

Calories from fat are what percent of total calories?

$288 = x \cdot 480$

Step 3: We can now use the appropriate form of the percent equation to find the percent:

$$x = \frac{a}{b} = \frac{\text{part}}{\text{whole}} = \frac{288}{480} = 0.6$$

We found the percent of fat in decimal form.

Step 4: Convert the decimal into a percent:

$0.6 = 60\%$

Solution: Fat constitutes 60% of the total calories.

Percent of Change

When both a beginning and ending amount of some quantity are given and we are asked to find the percent of change, we need to use the following formula:

$$\frac{\text{amount of increase or decrease}}{\text{original amount}} = \text{percent of change}$$

According to this formula, first find out how much the increase or decrease was. Then divide that by the original amount given by the problem. You will have your answer as a fraction or a decimal. Change that into a percent.

DEAD ENDS

You can't change a fraction to a percent directly. First change a fraction into a decimal, and then change the decimal into a percent.

Sample Problem 13

The enrollment at a local university increased from 12,000 students to 15,000 students over a period of 5 years. What is the percent of change in enrollment?

Step 1: We begin by finding how much the increase was:

15,000 – 12,000 = 3,000 students

We found that the amount of increase was 3,000 students.

Step 2: Now we divide that number by the original enrollment, which was 12,000 students, to find the percent of change:

$$\frac{3,000}{12,000} = \frac{1}{4}$$

We found the percent of increase as a fraction.

Step 3: To find the increase in a percent form, first change the fraction $\frac{1}{4}$ into a decimal:

$$\frac{1}{4} = \frac{1 \cdot 25}{4 \cdot 25} = \frac{25}{100} = 0.25$$

Change the decimal 0.25 into a percent form:

0.25 = 25%

Solution: The university enrollment increased by 25% over 5 years.

A WORD OF ADVICE

Change a fraction to a decimal in one of two ways:

- Change the denominator to a power of 10: $\frac{3}{5} = \frac{3 \cdot 2}{5 \cdot 2} = \frac{6}{10} = 0.6$
- Divide: 5:8 = 0.625

Sample Problem 14

My real estate agent told me that, due to the economic crisis, my house has gone from being worth $360,000 to being worth $315,000. What is the percent decrease in the value of my house?

Step 1: Find out how much the decrease is:

$360,000 – $315,000 = $45,000

We found that the amount of decrease is $45,000.

Step 2: Divide that number by the original price to find the percent of decrease:

$$\frac{\$45,000}{\$360,000} = 0.125$$

We found the percent of decrease in a decimal form.

Step 3: Change the decimal into a percent:

0.125 = 12.5%

Solution: The percent decrease in the value of my house is 12.5%.

Practice Problems

Problem 1: What number is 2% of 150?

Problem 2: 25% of what is 85?

Problem 3: Sheila and Todd work for a show in Las Vegas. If they paid $75,500 as their taxes, which were 25%, what was the total amount of money they earned?

Problem 4: Maria works at the veterinary clinic. If 28% of her patients last week were cats and she attended to 35 cats, how many animals did she see last week?

Problem 5: The number of points Sara has on her math tests is 480 out of a possible 640 points. What is her percentage score?

Problem 6: A famous deejay from Richmond played 175 songs at the Red Skies Party. If he played 35 European songs, what percent of European music did he play?

Problem 7: The selling price of a house dropped from $400,000 to $390,000. What is the percent change of the price drop?

The Least You Need to Know

- The basis of many percent problems is the formula $a = p\% \cdot b$.
- Only three types of percent problems exist: finding a part of the whole, finding the whole when the percent is given, and finding what percent one number is of another. All three types can be solved using the appropriate form of the basic formula.

- To use the basic percent formula, first convert a percent into its decimal form.
- When finding a percent, as a final step, convert a decimal into a percent by moving the decimal point two places to the right and writing the percent symbol.

Number, Age, and Work Problems

This part deals with the most common topics for word problems: numbers, age, and work.

First we discuss number problems. Humans have been fascinated with numbers since the dawn of civilization. Discovering the properties and rules that govern them was a difficult endeavor for professionals and an enjoyable pastime for many others. This chapter introduces you to general techniques and strategies for number problems and teaches you to find an unknown number, consecutive integers, or digits in a multiple digit number.

Later, we deal with age problems and learn how to use tables to organize the problem's given data and facilitate the writing of the problem's equation.

In the last chapter of this part, we explore the so-called "work formula" that enables you to calculate how much faster two persons or pipes can do the job in comparison to a single person or pipe.

The Numbers Game

In This Chapter

- Recording number relationships with algebraic expressions
- Finding even and odd integers
- Finding the smallest and largest numbers
- Interchanging the digits

Number puzzles are as old as the history of civilization itself. Since ancient times, people have been fascinated with numbers. Exploration of numbers and manipulations with them allowed people to discover the numbers' properties, unexpected relationships, and even new types of numbers.

In this chapter, we will learn how to solve some number puzzles and find numbers when their relationships with other numbers are given. Later, we will discuss how to find missing consecutive integers and explore what happens to a number when its digits are interchanged.

Find Numbers Given Their Relationships

Number problems can involve one, two, or even more numbers. These numbers can be whole numbers, integers, or even fractions. When a single number is involved, we need to perform some operations with it to find its value.

Sample Problem 1

One third of a number is 3 less than one half of the number. Find the number.

Step 1: Let x represent the number. Next, we express each piece of information about the number as an algebraic expression. One third can be expressed as $\frac{x}{3}$, and one half can be presented as $\frac{x}{2}$. Three less than one half can be expressed as $\frac{x}{2} - 3$.

Step 2: Write a word expression about the number and its mathematical equivalent:

One third is 3 less than one half.

$$\frac{x}{3} = \frac{x}{2} - 3$$

DEAD ENDS

Don't try to write an equation for number problems right away. First express each small bit of information mathematically and then connect them into an equation.

Step 3: Write the equation in a compact way and solve by multiplying each term by the least common denominator (LCD, which is 6) to eliminate fractions:

$$\frac{x}{3} = \frac{x}{2} - 3$$

$$\frac{\overset{2}{\cancel{6}}x}{\underset{1}{\cancel{3}}} = \frac{\overset{3}{\cancel{6}}x}{\underset{1}{\cancel{2}}} - 6 \cdot 3$$

Multiply the remaining factors:

$$2x = 3x - 18$$

Isolate the variable and collect like terms:

$$2x - 3x = -18$$

$$-x = -18$$

Divide both sides by –1:

$$x = 18$$

Solution: The number is 18.

Sample Problem 2

Fifteen less than 5 times a certain number equals 4 times the number plus 16. Find the number.

Step 1: Let x represent the number. Let's now "translate" each piece of information into algebra language. Five times a certain number becomes $5x$. Fifteen less than $5x$ can be expressed as $5x - 15$. Four times the number is $4x$, and 4 times plus 16 is $4x + 16$.

Step 2: Write a word expression and its algebra equivalent:

15 less than 5 times a number equals 4 times the number plus 16.

$5x - 15 = 4x + 16$

Step 3: Isolate the variable and collect like terms:

$5x - 4x = 16 + 15$

$x = 31$

Solution: The number is 31.

When more than one number is involved, some relationships between numbers are given. Numbers are compared, and this comparison allows us first to express the numbers' relationships in algebra language and, second, to write the equation.

Sample Problem 3

One number exceeds another number by 5, and the sum of the smaller number and twice the larger number is 61. Find the larger number.

Step 1: Since we know that the first number exceeds the other by 5, let that other number be represented by x; then the first number will be $x + 5$ (*exceeds* means "larger" in mathematical language). Twice the larger number ($x + 5$) can be expressed as $2(x + 5)$.

Step 2: Let's write a word expression and its algebra equivalent:

The sum of the smaller number and twice the larger number is 61.

$x + 2(x + 5) = 61$

Step 3: Distribute the right side of the equation:

$x + 2x + 10 = 61$

Isolate the variable and collect like terms:

$x + 2x = 61 - 10$

$3x = 51$

Divide both sides by 3:

$x = 17$

We found that the smaller number is 17.

Step 4: To find the larger number, add 5 to the smaller number, since the larger number exceeds the smaller by 5:

$17 + 5 = 22$

Solution: The larger number is 22.

A WORD OF ADVICE

When you are asked to find two numbers, let x represent the number that you know the least about. For example, if one number is 5 larger than the other, let the other number be x, since you know that the first number is 5 larger. Then the first number will be $x + 5$ and the second number will be x.

Sample Problem 4

One number is $\frac{4}{5}$ of another number. The sum of the two numbers is 81. Find the two numbers.

Step 1: Let the other number be represented by x, since we don't know anything about this number. The first number is $\frac{4}{5}$ of that number. The word *of* means multiplication in this case, so we can express the first number as $\frac{4x}{5}$.

Step 2: We are now ready to write a word expression and its algebraic equivalent:

The sum of two numbers is 81.

$x + \dfrac{4x}{5} = 81$

Step 3: Multiply each term by 5 and reduce by common factors to eliminate the fraction:

$$5x + \frac{\overset{1}{\cancel{5}} \cdot 4x}{\underset{1}{\cancel{5}}} = 5 \cdot 81$$

Multiply the remaining factors:

$5x + 4x = 405$

Collect like terms:

$9x = 405$

Divide both sides by 9:

$x = 45$

We found that the second number is 45.

Step 4: To find the first number, we need to multiply this number by $\frac{4}{5}$:

$$\frac{4}{5} \cdot 45 = \frac{4 \cdot 45}{5} = \frac{4 \cdot \overset{9}{\cancel{45}}}{\underset{1}{\cancel{5}}} = 36$$

We found that the first number is 36.

Solution: The two numbers are 36 and 45.

Sample Problem 5

The difference of two numbers is 7. If 3 times the larger number is 16 more than 4 times the smaller number, find the two numbers.

Step 1: Let the smaller number be x; then the larger number will be $x + 7$, since the difference between the two numbers is 7. The expression "3 times the larger number" translates into $3(x + 7)$, and "4 times the smaller" can be expressed as $4x$.

Step 2: Next, to equalize the two expressions $3(x + 7)$, which is 3 times the larger number, and $4x$, which is 4 times the smaller number, we need to add 16 to the second expression, since the first expression is 16 more.

Step 3: Now we are ready to write a word expression and its mathematical equivalent for the problem:

3 times the larger number is 16 more than 4 times the smaller.

$3(x + 7) = 16 + 4x$

Step 4: To solve the equation, distribute the left side and isolate the variable:

$3x + 21 = 16 + 4x$

$3x - 4x = -21 + 16$

Collect like terms:

$-x = -5$

Divide both sides by -1:

$x = 5$

We found that the smaller number is 5.

Step 5: To find the larger number, we substitute 5 into the expression for the larger number from Step 1:

$x + 7 = 5 + 7 = 12$

We found that the larger number is 12.

Solution: We found that the two numbers are 5 and 12.

WORTHY TO KNOW

Myriad is a classical Greek name for the number 10,000. Myriad myriads, or one hundred million, was the largest named number in the Bible. By using a system of numbers based on powers of the myriad, Archimedes concludes in his book *The Sand Reckoner* that the number of grains of sand required to fill the universe is 8×10^{63} in modern notation.

When three numbers are involved in a problem, the routine is the same: represent the number that we know the least about with x and express other numbers using given relationships.

Sample Problem 6

The sum of three numbers is 96. The first is 3 times the second number, and the third is 4 times the result obtained by subtracting 6 from the second number. Find the largest number.

Step 1: Since we know the least about the second number, let's represent it by x. Then the first number, which is 3 times the second, can be expressed as $3x$. The third number is a bit more complicated. First, we get the algebra equivalent of the words "the result obtained by subtracting 6 from the second number." This simply means that we need to subtract 6 from the second number: $x - 6$. And "4 times this result" can be expressed as $4(x - 6)$. So the third number is $4(x - 6)$.

Step 2: We assigned all variables and are ready to write the word expression and its algebraic equivalent:

The sum of three numbers is 96.

$3x + x + 4(x - 6) = 96$

Step 3: To solve the equation, distribute the left side:

$3x + x + 4x - 24 = 96$

Isolate the variable and collect like terms:

$3x + x + 4x = 24 + 96$

$8x = 120$

Divide both sides by 8:

$x = 15$

We found that the second number is 15.

Step 4: To find the first number, we need to multiply the second number by 3:

$3x = 3(15) = 45$

We found that the first number is 45. But we still don't know if this is the largest number, so we need to find the third number.

Step 5: To find the third number, we need to plug 15 instead of x into the third number expression:

$4(x - 6) = 4(15 - 6) = 60 - 24 = 36$

The third number is 36. So the first number is the largest.

Solution: The largest number is 45.

Consecutive Integers

This section of the chapter deals with *consecutive integers*. It is customary to represent the first integer in a sequence as n. The second integer then will be $n + 1$, the third will be $n + 2$, and so forth. The next step is to write the relationships between the consecutive integers in algebraic language. The most common operation performed in this type of problem is addition.

> **DEFINITION**
>
> **Consecutive integers** are integers that follow in sequence, each number being 1 more than the previous number—for example: 12, 13, 14, and so forth. Consecutive integers can be generally represented by $n, n + 1, n + 2, n + 3$, and so forth, where n is any integer.

Sample Problem 7

The sum of three consecutive integers is 105. What are the numbers?

Step 1: Let the first integer be represented by n; then the second can be expressed as $n + 1$ and the third as $n + 2$.

Step 2: Write the word expression and its algebraic equivalent:

The sum of three consecutive integers is 105.

$n + (n + 1) + (n + 2) = 105$

The parentheses aren't really necessary. We use them to emphasize that there are three different numbers in the sum.

Step 3: Open parentheses:

$n + (n +1) + (n + 2) = 105$

$n + n + 1 + n + 2 = 105$

Isolate the variable and collect like terms:

$n + n + n + = 105 - 2 - 1$

$3n = 102$

Divide both sides by 3:

$n = 34$

We found that the first integer is 34.

Step 4: To find other numbers, we substitute 34 into other number expressions from Step 1:

$n + 1 = 34 + 1 = 35$

$n + 2 = 34 + 2 = 36$

We found that the other two numbers are 35 and 36.

Solution: The three consecutive numbers are 34, 35, and 36.

Sample Problem 8

The largest of six consecutive integers is twice the smallest. Find the largest number.

Step 1: Let the smallest number be n; then all the following integers can be presented as $n + 1$, $n + 2$, $n + 3$, $n + 4$, and $n + 5$ (the largest). "Twice the smallest" can be expressed as $2n$.

Step 2: We are ready to write a sentence and its algebraic equivalent:

The largest integer is twice the smallest.

$n + 5 = 2n$

Step 3: Isolate the variable:

$n - 2n = -5$

Collect like terms and divide both sides by –1:

$-n = -5$

$n = 5$

We found that the smallest integer is 5.

Step 4: To find the largest integer, we substitute 5 into the expression for the largest integer from Step 1:

$n + 5 = 5 + 5 = 10$

Solution: The largest integer is 10.

Many number problems involve *even and odd consecutive integers*. They have a common difference of two between each term.

DEFINITION

If we start with an even integer and each number in the sequence is 2 more than the previous number, then we have **consecutive even integers**—for example: 4, 6, 8, 10 ….

If we start with an odd number and each number in the sequence is 2 more than the previous number, then we have **consecutive odd integers**—for example: 7, 9, 11, 13 ….

Sample Problem 9

Find three consecutive even integers such that the sum of twice the first and 3 times the third is 142.

Step 1: Let's first assign variables. The first number will be represented by n; then the second even integer would be $n + 2$, and the third number would be $n + 4$. Note that the difference between the numbers is always 2, since we are interested only in even numbers. "Twice the first" can be expressed as $2n$, and "3 times the third" can be expressed by multiplying the third integer $(n + 4)$ by 3: $3(n + 4)$. "Sum" means that we need to add these two numbers.

Step 2: We are ready to write a sentence and its algebraic equivalent:

The sum of twice the first and 3 times the third is 142.

$2n + 3(n + 4) = 142$

Step 3: Distribute the left side:

$2n + 3n + 12 = 142$

Isolate the variable and collect like terms:

$2n + 3n = 142 - 12$

$5n = 130$

Divide both sides by 5:

$n = 26$

We found that the smallest even integer is 26.

Step 4: To find the other two integers, we substitute 26 into the expressions for the other two numbers from Step 1:

$n + 2 = 26 + 2 = 28$

$n + 4 = 26 + 4 = 30$

Solution: The three even consecutive integers are 26, 28, and 30.

Sample Problem 10

Find three odd consecutive integers such that twice the sum of the first and the second is 5 more than 3 times the third.

Step 1: Let n represent the first odd integer; then the second can be expressed as $n + 2$, and the third as $n + 4$. The word expression "twice the sum of" means that first we need to find the sum and then multiply it by two. The sum of the first and the second integers will be $n + (n + 2)$, and twice that sum can be expressed as $2[n + (n + 2)]$. Finally, "3 times the third" can be expressed as $3(n + 4)$. Since twice the sum of the first two integers is 5 more than 3 times the third, we need to add 5 to the latter to equalize both parts.

Step 2: Write the sentence and its mathematical equivalent:

Twice the sum of the first and the second is 5 more than 3 times the third.

$2[n + (n + 2)] = 5 + 3(n + 4)$

Step 3: Distribute both sides:

$2[n + (n + 2)] = 5 + 3(n + 4)$

$2n + 2n + 4 = 5 + 3n + 12$

Isolate the variable:

$2n + 2n - 3n = -4 + 5 + 12$

Collect like terms:

$n = 13$

We found that the first odd integer is 13.

Step 4: To find the other two odd integers, let's substitute 13 into their expressions from Step 1:

$n + 2 = 13 + 2 = 15$

$n + 4 = 13 + 4 = 17$

Solution: Three consecutive odd integers are 13, 15, and 17.

> **A WORD OF ADVICE**
>
> Use grouping symbols to express consecutive numbers and relationships between them. For example, write the expression 2 times the sum of two consecutive numbers as $2[n + (n + 1)]$.

Digit Problems

Digit problems involve individual digits and how the digits relate to each other. To some extent, these problems are very similar to integer problems, except for the fact that digits are only between 0 and 9, inclusive.

Sample Problem 11

The tens digit of a two-digit number is 3 times the ones digit. The sum of the digits in the number is 12. Find the number.

Step 1: Assign variables and let x represent the ones digit. Then the tens digit, which is 3 times the ones, can be expressed as $3x$. The problem also gives us the sum of the digits, so we need to add them.

Step 2: Write the sentence and its mathematical equivalent:

The sum of the digits in the number is 12.

$x + 3x = 12$

Step 3: Collect like terms on the left:

$x + 3x = 12$

$4x = 12$

Divide both sides by 4:

$x = 3$

We found that the ones digit is 3.

Step 4: To find the tens digit, we multiply the ones digit by 3:

$3x = 3(3) = 9$

We found that the tens digit is 9.

We now know both digits, so we can write the number: 93.

Solution: The number is 93.

If the problem involves a comparison between the sum of the number's digits and the number itself, we need to use the *expanded form* for the number (multiply the digits by the place value of the digits).

> **DEFINITION**
>
> The value of each digit in a whole number is shown by writing the number in **expanded form.**
>
> Standard form: 93,756
>
> Expanded form: 10,000(9) + 1,000(3) + 100(7) + 10(5) + 1(6)

Sample Problem 12

The tens digit of a two-digit number is 4 more than the ones digit. If the number is 12 less than 8 times the sum of the digits, find the number.

Step 1: Let x represent the ones digit; then the tens digit can be expressed as $x + 4$, since it is 4 more than the ones digit. The sum of the digits will be $x + (x + 4)$, and 8 times the sum will be $8[x + (x + 4)]$.

Step 2: Next, we need to express the number itself. We use its expanded form and the expressions for digits we found in Step 1:

The number in its expanded form: $10(x + 4) + 1(x)$.

Since the number is 12 less than 8 times the sum of the digits, in order to equalize both, we need to subtract 12 from the sum.

Step 3: Write the sentence and its algebraic equivalent:

The number is 12 less than 8 times the sum of the digits.

$10(x + 4) + 1(x) = 8[x + (x + 4)] - 12$

Step 4: Distribute on both sides:

$10(x + 4) + 1(x) = 8[x + (x + 4)] - 12$

$10x + 40 + x = 8x + 8x + 32 - 12$

Isolate the variable:

$10x + x - 8x - 8x = -40 + 32 - 12$

Collect like terms:

$-5x = -20$

Divide both sides by −5:

$x = 4$

We found that the ones digit is 4.

Step 5: To find the tens digit, we use the expression for it from Step 1 and substitute 4 for the ones digit:

$x + 4 = 4 + 4 = 8$

We found that the tens digit is 8. Then the number is 84.

Solution: The number is 84.

When numbers in digit problems are three or more digits, it is convenient to use a chart to record the digits.

Sample Problem 13

In a certain three-digit number, the hundreds digit is 2 less then the tens digit, and the sum of the digits is 15. If the ones and hundreds digits are interchanged, the number is increased by 396. Find the original number.

Step 1: Let's first sort things out with the original number. Since the relationships are given between the tens and hundreds digits and we know the least about the tens digit, let's denote it by x. Then the hundreds digit will be $x - 2$, since it is 2 less. To find the ones digit, we use the fact that the sum of all digits is 15, so we subtract the sum of the other two digits $[x + (x - 2)]$ from 15 to find the ones digit:

$15 - [x + (x - 2)] = 15 - x - x + 2 = 17 - 2x$

Step 2: Let's make a chart and record the digits of the original number:

hundreds digit	tens digit	ones digit
$x - 2$	x	$17 - 2x$

Step 3: We write the original number using its expanded form and digits from Step 2 as this:

$100(x - 2) + 10x + 1(17 - 2x)$

Step 4: Let's now deal with the new number. Since we are interchanging the ones and hundreds digits, the chart for this new number is this:

hundreds digit	tens digit	ones digit
$17 - 2x$	x	$x - 2$

Step 5: We write the new number in its expanded form using the expressions for digits from Step 4 as this:

$100(17 - 2x) + 10x + 1(x - 2)$

Step 6: Since the new number was increased by 396 in the interchanging process, we need to subtract 396 from this number to equalize both numbers. We write the equation that equalizes both numbers using the expanded forms for both numbers and the fact that the new number is larger by 396:

Original number equals the new number minus 396.

$100(x - 2) + 10x + 1(17 - 2x) = 100(17 - 2x) + 10x + 1(x - 2) - 396$

Step 7: Distribute on both sides:

$100x - 200 + 10x + 17 - 2x = 1,700 - 200x + 10x + x - 2 - 396$

Isolate the variable:

$100x + 10x - 2x + 200x - 10x - x = 200 - 17 + 1,700 - 2 - 396$

Collect like terms:

$297x = 1,485$

Divide both sides by 297:

$x = 5$

We found that the tens digit is 5.

Step 8: To find the other digits, substitute 5 into expressions for other digits from Step 1:

Hundreds digit: $x - 2 = 5 - 2 = 3$

Ones digit: $17 - 2x = 17 - 2(5) = 17 - 10 = 7$

We know all the digits, so we can write the original number: 357.

Solution: The original number is 357.

Sometimes the problem does not provide the relationship between the digits. In this case, we have two options. The first option is to represent the first digit as x and then express the other digit using given values. For example, if the sum of two digits is given, then the second digit can be expressed as (sum $- x$). The second choice is to use two variables and, subsequently, solve a system of two linear equations.

Sample Problem 14

The sum of the digits of a two-digit number is 16. The value of the number is 20 less than 13 times the tens digit. Find the number.

Step 1: Since we know the sum of the digits but not the relationship between the units and tens digits, we use two variables for this problem. Let x represent the tens digit and y represent the ones digit. Then we can state the fact that the sum of the digits is 16 mathematically:

$x + y = 16$

Step 2: Next, we need to find the value of the number, since this value is compared with 13 times the tens digit. Using the expanded form for the number, we can write that the value of the number is this:

$10x + 1y$

WORTHY TO KNOW

Indian mathematicians around 500 C.E. invented the base 10 system that had unique symbols for the numbers 1 through 9, used a place value notation, and used a zero. This is the system that evolved into the number system we use today.

Step 3: 13 times the tens digit can be expressed as $13x$. Since the number value is 20 less than 13 times the tens digit, to equalize both, we need to subtract 20 from 13 times the tens digit ($13x - 20$).

Step 4: We can now write the sentence and its mathematical equivalent:

The value of the number is 20 less than 13 times the tens digit.

$10x + 1y = 13x - 20$

Step 5: Using two equations from Step 1 and Step 4, we obtain the system of two linear equations:

$$\begin{cases} x + y = 16 \\ 10x + y = 13x - 20 \end{cases}$$

We solve the system using the substitution method. Revise the first equation:

$y = 16 - x$

Substitute this expression into the other equation and solve for x:

$10x + (16 - x) = 13x - 20$

Isolate the variable:

$10x - x - 13x = -16 - 20$

Collect like terms:

$-4x = -36$

Divide both sides by -4:

$x = 9$

We found that the tens digit is 9.

Step 6: To find the units digit, substitute 9 into the revised equation from Step 5:

$y = 16 - x = 16 - 9 = 7$

We found that the units digit is 7. We know both digits and can write the number: 97.

Solution: The number is 97.

Practice Problems

Problem 1: 8 times the number is 14 more than 7 times the number. Find the number.

Problem 2: The sum of three numbers is 152. The second number is 3 times the first, and the third is 4 times the first. Find the numbers.

Problem 3: Find five consecutive numbers whose sum is 65.

Problem 4: The sum of the least and greatest of four consecutive integers is 49. What is the third integer?

Problem 5: Three times one odd integer is 11 more than 2 times the next even integer. Find the integers.

Problem 6: Find three consecutive odd integers such that 3 times the sum of the first and the second is 3 more than 5 times the third.

Problem 7: In a certain three-digit number, the units digit is 3 more than the tens digit, and the sum of the digits is 11. If the units and hundreds digits are interchanged, the number increases by 99. Find the original number.

The Least You Need to Know

- A variable always represents a number or a digit.
- Addition is the most common operation in number word problems.
- Assign x to represent the number or digit you know the least about.
- It is useful to use a chart to record digits in digit problems.

Age Is Relative

Age word problems are problems that ask for one person's age or several people's ages. The goal is to retrieve the answer by setting up and solving algebraic equations. These types of word problems have been known for more than 2,000 years. I suspect that they were created by a smart lady mathematician who invented this brilliant way of not directly answering the question about her age. I'm sure that very few people at this time were able to solve these problems, since the language of algebra was in the early stages of development. In modern times, we have the full power of algebra at our disposal, but it doesn't seem that age problems have become any easier to solve. This chapter is all about determining age based on given information.

Problems That Involve a Single Person

This type of age problem deals with finding one person's age. The problems compare the person's age in the past, present, and future. In my opinion, these are the easiest type of age problems. Let's warm up by solving them.

Sample Problem 1

In 5 years, Anna will be 2 times older than she is today. How old is Anna today?

Step 1: Since the problem asks you to find Anna's age today, let's just say that x is Anna's age today. If Anna's age today is x, then in 5 years, her age will be $x + 5$.

Step 2: Also, we know that in 5 years she will grow to be "2 times older than she is now." If Anna is x years old today, then "2 times older" can be expressed as $2x$.

Step 3: Using variables and expressions from Steps 1 and 2, we translate the following statement into algebraic language:

In 5 years, Anna will be 2 times older than she is now.

$x + 5 = 2x$

Isolate the variable:

$x - 2x = -5$

Collect like terms:

$-x = -5$

Divide both sides by −1:

$x = 5$

Solution: Anna is 5 years old today.

DEAD ENDS

When you obtain the answer to the equation, always check it by plugging it back into the equation and make sure that it all makes sense. You know that people rarely live up to more than 120 years. If your answer is some crazy age, check the setup of your equation and its solution.

Sample Problem 2

Ten years ago, Ben's age was half the age he will be in 16 years. How old is Ben now?

Step 1: Let's denote Ben's age today as x. Then "10 years ago" translates into $x - 10$.

Step 2: We can also state that, since Ben is x years old now, "in 16 years," he will be x + 16. Half of that is:

$$\frac{1}{2}(x+16)$$

Step 3: Now we're ready to write out the equation:

Ten years ago, Ben's age was half the age he will be in 16 years.

$$x - 10 = \frac{1}{2}(x+16)$$

Multiply by 2 to eliminate fractions:

$$2(x - 10) = x + 16$$

Open parentheses on the left side:

$$2x - 20 = x + 16$$

Isolate the variable and collect like terms:

$$2x - x = 20 + 16$$

$$x = 36$$

Solution: Ben is 36 years old now.

A WORD OF ADVICE

As we know, words like *was, is,* and *will be* are translated into an equals sign (=). If in a problem the phrase "half the age" is located to the right of one of these words, the fraction $\frac{1}{2}$ should also be placed to the right of the equals sign.

Let's do the last problem in this section.

Sample Problem 3

In 10 years, Mary will be 3 times plus 2 as old as she is now. How old is Mary today?

Step 1: Let x represent Mary's age today. Then "in 10 years" can be expressed as x + 10, and "3 times plus 2 as old as she is now" translates into $3x + 2$.

Step 2: Using the variables and expressions from Step 1, we can now write the equation:

In 10 years, Mary will be 3 times plus 2 as old as she is now.

$x + 10 = 3x + 2$

Isolate the variable:

$x - 3x = -10 + 2$

Collect like terms:

$-2x = -8$

Divide both sides by -2:

$x = 4$

Solution: Mary is 4 years old now.

Problems That Involve Multiple People

Let's now discuss age problems that involve more than one person. In these types of problems, besides comparing the ages of different people, we have to jump among the present, past, and future to find out their age. It gets interesting and complicated at the same time. Using a chart is useful for recording people's ages at current time, in the past, or in the future so you don't get completely lost.

Sample Problem 4

My friend Sam is now twice as old as his sister. Four years ago he was 3 times as old as his sister. How old is my friend Sam now?

Step 1: Let x represent Sam's little sister's age today. Since Sam is now twice as old as his sister, his age now can be expressed as $2x$.

Step 2: If the sister is x years old now, then 4 years ago, the sister's age was $x - 4$. Accordingly, Sam's age 4 years ago was $2x - 4$ (his current age minus 4).

Step 3: Using the variables and values we found in Steps 1 and 2, create a chart to organize the problem. The first column shows the people involved. The second column lists the people's current ages. The third column displays their ages 4 years ago.

	Now	4 Years Ago
Sister	x	$x - 4$
Sam	$2x$	$2x - 4$

Step 4: Now we can set up an equation using the third column of the chart. To translate the phrase "3 times as old as his sister," we need to multiply the sister's age 4 years ago by 3.

Four years ago, Sam was 3 times as old as his sister.

$2x - 4 = 3(x - 4)$

Distribute the right side:

$2x - 4 = 3x - 12$

Isolate the variable:

$2x - 3x = 4 - 12$

Collect like terms:

$-x = -8$

Divide both sides by -1:

$x = 8$

We found that the sister is 8 years old now. This is the answer to the equation, but not to the problem's question.

Step 5: We can find Sam's age by substituting the sister's age into the expression for Sam's current age from Step 1:

$2x = 2(8) = 16$

Solution: Sam is 16 years old now.

DEAD ENDS

You might set up and solve the equation correctly, but the answer to the problem might still be quite different. This happens when you answer the wrong question. For convenience purposes, sometimes we denote by variable x the age that's not the question at the end of the problem. Always read the problem's question carefully and answer it.

Let's do more examples now.

Sample Problem 5

Olga is 16 years younger than Peter. Five years ago, Olga was $\frac{2}{3}$ as old as Peter. How old are they now?

Step 1: Olga's age is compared to Peter's, so let's say that his age is x. Then Olga is $x - 16$ years old now, since she is 16 years younger.

Step 2: Five years ago, Peter's age was $x - 5$. To find Olga's age five years ago, we need to subtract 5 from her current age of $(x - 16)$: $(x - 16) - 5 = x - 16 - 5 = x - 21$. We found that Olga's age 5 years ago was $x - 21$.

Step 3: Using the variables and values we found in Steps 1 and 2, create a chart to organize the problem. The first column shows the names of the people. Their age now is listed in the second column. The third column displays their age 5 years ago.

	Now	5 Years Ago
Peter	x	$x - 5$
Olga	$x - 16$	$x - 21$

Step 4: Using the information from the third column and the fact that, 5 years ago, Olga's age was $\frac{2}{3}$ as old as Peter, we can set up the problem's equation:

Five years ago, Olga was $\frac{2}{3}$ as old as Peter.

$x - 21 = \frac{2}{3}(x - 5)$

Multiply both sides by 3 to eliminate fractions:

$3(x - 21) = 2(x - 5)$

Distribute on both sides:

$3x - 63 = 2x - 10$

Isolate the variable:

$3x - 2x = 63 - 10$

Collect like terms on both sides:

$x = 53$

We found that Peter is 53 years old now.

Step 5: To find Olga's age, let's substitute x into the expression for Olga's age from Step 1:

$x - 16 = 53 - 16 = 37$

We found that Olga is 37 years old now.

Solution: Peter is 53 and Olga is 37 years old.

Sometimes more than two people are involved in age problems. Let's do one of these examples.

Sample Problem 6

Uncle Jim is 3 times older than his son Bill, and Bill is 5 times as old as his sister Carry. In 2 years, the sum of their ages will be 69. How old is Bill now?

Step 1: Since Carry is the youngest, it is convenient to let x stand for her current age. Then Bill's age can be expressed as $5x$, since Bill is 5 times older than Carry. Uncle Jim's age can be $3(5x) = 15x$, since he is 3 times older than Bill.

Step 2: In 2 years, Carry will be $x + 2$ years old, Bill will be $5x + 2$ years old, and Uncle Jim, who is $15x$ years old now, will be $15x + 2$ years old.

Step 3: Using the variables and values we found in Steps 1 and 2, create a chart to organize the problem. Since we have three people, our chart has one additional row for the third person. The first column shows the people involved. The second column lists the people's current ages. The third column displays their ages 2 years from now.

	Now	**In 2 Years**
Bill	$5x$	$5x + 2$
Uncle Jim	$15x$	$15x + 2$
Carry	x	$x + 2$

Step 4: Now we're ready to write the equation using information from the third column and the fact that their combined age will be 69 in 2 years:

In 2 years, the sum will be 69.

$(x + 2) + (5x + 2) + (15x + 2) = 69$

Open parentheses and combine like terms on the left side:

$x + 2 + 5x + 2 + 15x + 2 = 69$

$21x + 6 = 69$

Isolate the variable and collect like terms:

$21x = 69 - 6$

$21x = 63$

Divide both sides by 21:

$x = 3$

We found that Carry is 3 years old now.

Step 5: To find Bill's age now, we need to substitute 3 into the expression for Bill's age from Step 1:

$5x = 5(3) = 15$

We found that Bill is 15 years old now.

Solution: Bill is 15 years old now.

A WORD OF ADVICE

The question may arise whether you have a single way to choose whose age you denote as *x*. The answer to this question is "no." You have several choices, but usually one of them offers the easiest way to solve the problem. Regardless of whose age you denote as *x*, the final answers to the problem (the ages of people) will be the same.

Sample Problem 7

Dad is 3 times as old as his daughter Sara. If the sum of their ages is 25 more than twice Sara's age plus 5, how old will Sara be in 7 years?

Step 1: Since we're not traveling through time as much as in previous problems, we don't really need our chart here. Instead, we need to break down the problem into component parts and write them down as algebraic expressions. Let's start by letting Sara's age be *x*. Then her dad's age is $3x$, since he is 3 times as old as Sara. Twice Sara's age can be expressed as $2x$, and twice Sara's age plus 5 translates into $2x + 5$.

Step 2: We're now ready to set up our equation using the variables and values we found in Step 1:

The sum of their ages is 25 more than twice Sara's age plus 5.

$x + 3x = 25 + (2x + 5)$

Isolate the variable:

$x + 3x - 2x = 25 + 5$

Collect like terms:

$2x = 30$

Divide both sides by 2:

$x = 15$

We found that Sara is 15 years old now.

Step 3: To find Sara's age in 7 years, add 7 to her current age:

$15 + 7 = 22$

Solution: In 7 years, Sara will be 22 years old.

Now we do the last and most challenging problem in this section.

Sample Problem 8

Silvia's age is 3 years more than twice her brother's age. Seven years ago, her age was 1 more than 3 times her brother's age. How old will the siblings be in 5 years?

If you feel a little dizzy after reading the problem, I don't blame you—it sounds pretty confusing. Let's apply our proven strategy: break the problem into chunks, and write them down as algebraic expressions.

Step 1: It's convenient to let the brother's age be x. Clearly, "twice her brother's age" translates into $2x$. Then the phrase "Silvia's age is 3 years more than twice her brother's age" becomes $2x + 3$.

Step 2: Now let's go back to the past and deal with the siblings' age back then. Seven years ago, the little brother was $x - 7$ years old. Silvia was "her current age minus 7," or $(2x + 3) - 7 = 2x - 4$.

Step 3: Using the variables and values we found in Steps 1 and 2, create a chart to organize the problem. The first column shows the people involved. The second column lists the people's current ages. The third column displays their ages 7 years ago.

	Now	7 Years Ago
Brother	x	$x - 7$
Silvia	$2x + 3$	$2x - 4$

Step 4: We can now set up the equation using the siblings' ages as they were 7 years ago:

7 years ago, her age was 1 more than 3 times her brother's age.

$2x - 4 = 1 + 3(x - 7)$

Distribute and collect like terms on the right side:

$2x - 4 = 1 + 3x - 21$

$2x - 4 = 3x - 20$

Isolate the variable and collect like terms:

$2x - 3x = -20 + 4$

$-x = -16$

Divide both sides by –1:

$x = 16$

We found that the little brother is 16 years old now.

Step 5: To find Silvia's age, substitute 16 into the expression for her age from Step 1:

$2x + 3 = 2(16) + 3 = 32 + 3 = 35$

We found that Silvia is 35 years old now.

To find people's ages in 5 years, add 5 to their current ages:

$16 + 5 = 21$

$35 + 5 = 40$

Solution: In 5 years, Silvia will be 40 and her brother will be 21.

Problems That Contain More Fractions

Discussions with my students led me to believe that the more fractions any problem has, the more complicated and unattractive it looks to the students. It seems that fractions have this mysterious power to convert any cute age problem into a monstrous one that students feel helpless to conquer. Let's do several of them. You might find that they're not as frightening as they seem upon the first glance.

Sample Problem 9

Two years ago, Sofia was $\frac{1}{4}$ as old as her mom. Fourteen years from now, she will be $\frac{5}{12}$ as old as her mom. How old is Sofia now?

Step 1: To say that Sofia is $\frac{1}{4}$ as old as her mom is the same as saying that her mom is 4 times older than Sofia. For example, if the mom is 48 now, Sofia's age is $\frac{1}{4}(48)=12$. On the other hand, 4(12) = 48. Let's say that Sofia's current age is x. Two years ago, she was $x - 2$ and her mom was $4(x - 2)$, since two years ago, the mom was 4 times older than her daughter.

Today, the mom's age is equal to her age 2 years ago plus 2:

$4(x - 2) + 2 = 4x - 8 + 2 = 4x - 6$

Step 2: In 14 years, Helen will be $x + 14$, since today she is x years old.

The mom will be the mom's age now $(4x - 6)$ plus 14 years:

$4x - 6 + 14 = 4x + 8$

Step 3: Using the variables and values we found in Steps 1 and 2, create a chart to organize the problem. Note that we have four columns in this problem, with an additional one for the present time. The first column shows the people involved. The second column shows people's ages 2 years ago. The third column lists the people's current ages. The fourth column displays their ages 14 years from now.

	2 Years Ago	Now	In 14 Years
Sofia	$x - 2$	x	$x + 14$
Mom	$4(x - 2)$	$4x - 6$	$4x + 8$

Step 4: Let's set up the equation:

14 years from now, Sofia will be $\dfrac{5}{12}$ as old as her mom.

$$x + 14 = \frac{5}{12}(4x + 8)$$

Multiply both sides by 12 to eliminate fractions:

$12(x + 14) = 5(4x + 8)$

Distribute on both sides and isolate the variable:

$12x + 168 = 20x + 40$

$12x - 20x = -168 + 40$

Collect like terms:

$-8x = -128$

Divide both sides by -8:

$x = 16$

Solution: Sofia is 16 years old now.

Sometimes age problems with fractions involve only one person, but they have a lot of fractions. One of the oldest problems of this type is about Diophantus. Very little is known about the life of this ancient mathematician who lived in Greece. All that we know about him was taken from an inscription on his tombstone in the form of an age problem. It reads:

> "Traveler! Here rest the ashes of Diophantus, and numbers can measure the length of his life. A sixth portion of it was a beautiful childhood. After another twelfth part of his life was over, beard covered his chin. Another seventh part of his life he spent in childless marriage. Five years then passed and he rejoiced in the birth of his son whom fate measured out a brilliant life only half of that of his father's. In deep grief the old man ended his life outliving his beloved son only by four years."

Let's now solve a problem that is similar to the one found on Diophantus's tombstone. This problem tells a true life story of my remote relative, Aunt Elka.

Sample Problem 10

My Aunt Elka was born in Russia and has been living a wonderful and fulfilling life. A twelfth portion of it was a joyful childhood. A sixth portion of her life she devoted to her studies and education. The next 7 years she spent working as a scientist. Then World War II began and ruined the lives and futures of many Russians. For $\frac{1}{48}$ th of her life, Aunt Elka lived in Leningrad, which was under siege by the Nazis, where thousands of people died of hunger and from enemy bombs. But she survived and met her future husband. They got married when the war ended. For half of her life they lived in Russia as a couple and were blessed with the birth of their children and grandchildren. Then Aunt Elka's family moved to the United States, where they have been living for 15 years now. How old is Aunt Elka?

Step 1: Let Aunt Elka's age be x. Then we can express different periods of her life as the following:

Childhood	$\dfrac{x}{12}$
Studies and education	$\dfrac{x}{6}$
Work as a scientist	7
Under siege	$\dfrac{x}{48}$
Life in Russia after marriage	$\dfrac{x}{2}$
Life in America	15

Now we can write down the equation:

The sum of all life portions is Elka's age.

$$\frac{x}{12}+\frac{x}{6}+7+\frac{x}{48}+\frac{x}{2}+15=x$$

Multiply both sides by 48 (which is the LCD) to eliminate fractions:

$$\frac{48\cdot x}{12}+\frac{48\cdot x}{6}+48\cdot 7+\frac{48\cdot x}{48}+\frac{48\cdot x}{2}+48\cdot 15=48\cdot x$$

Reduce by the common factors:

$$\frac{\overset{4}{\cancel{48}} \cdot x}{\underset{1}{\cancel{12}}} + \frac{\overset{8}{\cancel{48}} \cdot x}{\underset{1}{\cancel{6}}} + 336 + \frac{\overset{1}{\cancel{48}} \cdot x}{\underset{1}{\cancel{48}}} + \frac{\overset{24}{\cancel{48}} \cdot x}{\underset{1}{\cancel{2}}} + 720 = 48x$$

Multiply the remaining factors:

$4x + 8x + 336 + x + 24x + 720 = 48x$

Isolate the variable:

$4x + 8x + x + 24x - 48x = -720 - 336$

Collect like terms:

$-11x = -1,056$

Divide both sides by -11:

$x = 96$

Solution: Aunt Elka is 96 years old!

Here comes the last age problem we discuss in this chapter.

Sample Problem 11

Three years ago, Helen was $\frac{1}{6}$ as old as her mom. Eleven years from now, she will be $\frac{2}{5}$ as old as her mom. How old is Helen now?

Step 1: Since the problem asks for Helen's age now, let her current age be x. Three years ago, she was $x - 3$. Since Helen was $\frac{1}{6}$ as old as her mom, this is the same as saying that her mom was 6 times older than Helen. So the mom's age 3 years ago was $6(x - 3)$.

Step 2: Now let's deal with their ages 11 years from now. In 11 years, Helen will be $x + 11$ years old.

We know that the mom's age 3 years ago was $6(x - 3)$. Today, 3 years later, she is $6(x - 3) + 3 = 6x - 18 + 3 = 6x - 15$ years old.

In 11 years, she will be her current age plus 11: $(6x - 15) + 11 = 6x - 4$ years old.

Step 3: Using the variables and values we found in Steps 1 and 2, create a chart to organize the problem. Note that we have four columns in this problem, with an additional one for the present time. The first column shows the people involved. The second column shows their ages three years ago. The third column lists their current ages. The fourth column displays their ages 11 years from now.

	3 Years Ago	Now	In 11 Years
Helen	$x - 3$	x	$x + 11$
Mom	$6(x - 3)$	$6x - 15$	$6x - 4$

Step 4: We're now ready to set up an equation:

11 years from now, Helen will be $\frac{2}{5}$ as old as her mom.

$x + 11 = \frac{2}{5}(6x - 4)$

Multiply both sides by 5 to eliminate fractions:

$5(x + 11) = 2(6x - 4)$

Distribute on both sides:

$5x + 55 = 12x - 8$

Isolate the variable and collect like terms:

$5x - 12x = -55 - 8$

$-7x = -63$

Divide by -7:

$x = 9$

Solution: Helen is 9 years old now.

Hopefully, solving so many age problems makes you feel better about them and keeps you from feeling so intimidated. It's always beneficial to create your own problems for friends or relatives to solve. I know, this doesn't sound too appealing initially, but you may come to like it. For instance, for the next anniversary of your close friend, you might want to write the birthday card as an age problem similar to the one about Aunt Elka, highlighting the most interesting and memorable periods of his or her life. I'll bet the card would win the competition for being the most unique and unexpected.

Practice Problems

Problem 1: In 8 years, Catherine will be 2 times older than she is today. How old is Catherine today?

Problem 2: Four years ago, Bill's age was $\frac{1}{3}$ of the age he will be in 32 years. How old is he now?

Problem 3: Ashley is 10 years older than her little sister. Six years from now, she will be twice her sister's age. How old is Ashley now?

Problem 4: Jason is 10 years older than his brother. Eight years ago, the sum of their ages was 58. How old are they now?

Problem 5: Mark is 11 years older than twice Paul's age. Five years ago, the sum of their ages was 52. How old is Mark now?

Problem 6: A daughter is half as old as her mom. Twelve years ago, the daughter was $\frac{1}{3}$ as old as her mom. How old are they now?

Problem 7: Two years ago, Danny was $\frac{1}{5}$ as old as his uncle. Four years from now, Danny will be $\frac{1}{3}$ as old as his uncle. How old is the uncle now?

The Least You Need to Know

- Breaking age problems into component parts is helpful. For example, after denoting one's age as *x*, express others' ages in present or different times in terms of *x*.

- Often, we have more than one way to choose a variable, but some of these ways lead to fractional equations. Choosing the variable wisely makes the equation easier.

- Creating an age chart eliminates a lot of confusion. It helps to organize and record ages of people in the past, the present, and the future and makes it easy to write the problem's equation.

- Answering the right question is the key to success. For example, if you denote by *x* the age of a little sister, and the problem asks for the age of her brother who is 5 years older, solving an equation and providing the value for *x* does not answer the problem's question.

Working It

In This Chapter

- Finding rates
- Using the same units for time
- Dealing with fractions in work problems
- Exploring when to add or subtract rates

Work word problems involve different people (or machines) doing work together, but at different rates.

The main formula for work problems when two persons (or machines) are involved is:

$$\frac{1}{t_1} + \frac{1}{t_2} = \frac{1}{t_3}$$

t_1 is time taken by the first person.

t_2 is time taken by the second person.

t_3 is time taken by both working together.

The whole job is represented by the number 1, since the whole thing is always 100% = 1.

Also note that for the rest of this chapter when we mention two people mowing the lawn together, or two secretaries typing a document together, or three people doing something else together, it always means that they are doing the job simultaneously but using two or three pieces of equipment or machinery (one for each person).

Two People or Objects Working Together

The key to success with work problems is to think in terms of how much work each person or machine does in a certain unit of time. For example, when a person can do a job in 5 hours, then he can do $\frac{1}{5}$ of the job in 1 hour (the whole job is 1). This means that his rate is $\frac{1}{5}$ of the job per hour. If another man can do the same job in 3 hours, then he can do $\frac{1}{3}$ of the job in one hour, or his rate is $\frac{1}{3}$ of the job per hour. The sum of these rates $(\frac{1}{5}+\frac{1}{3})$ is the rate of two men working together. This is the basic idea in solving work problems.

Sample Problem 1

Tom can mow the lawn in 30 minutes, and his dad can mow the same lawn in 20 minutes. How long will it take them working together to mow the lawn?

Step 1: We don't know how much time it will take Tom and his dad to finish the job together, so let x indicate this time. Then their rate together will be $\frac{1}{x}$ of the job per minute.

Step 2: Tom can mow the lawn in 30 minutes, so his rate is $\frac{1}{30}$ of the job per minute. His dad can mow the lawn in 20 minutes, so his rate is $\frac{1}{20}$ of the job per minute.

WORTHY TO KNOW

Many work problems aren't realistic at all. (Since when does someone have a filling pipe to fill a tank of water and leave a draining pipe working at the same time?) Don't try to make sense of these problems—just try to learn the technique.

Step 3: We can set up the equation using the main work formula:

$$\frac{1}{t_1}+\frac{1}{t_2}=\frac{1}{t_3}$$

Plug in the rates from Steps 1 and 2:

$$\frac{1}{30}+\frac{1}{20}=\frac{1}{x}$$

To solve the equation, multiply each term by the least common denominator (LCD), which is $60x$:

$$\frac{60x \times 1}{30} + \frac{60x \times 1}{20} = \frac{60x}{x}$$

Reduce to eliminate the fractions:

$$\frac{\overset{2}{\cancel{60}}x}{\cancel{30}} + \frac{\overset{3}{\cancel{60}x}}{\cancel{20}} = \frac{60\cancel{x}}{\cancel{x}}$$

Multiply the remaining terms:

$2x + 3x = 60$

Collect terms on the left and divide both sides by 5:

$5x = 60$

$x = 12$ minutes

Solution: Tom and his dad together can finish mowing the lawn in 12 minutes.

A WORD OF ADVICE

Since the main formula for work problems contains rates, you have to think in terms of how much work each person, pipe, or machine does in a given unit of time.

Sample Problem 2

It takes Mary 3 hours to plant new flowers at their home garden. With the help of her little sister, she can finish the job in 2 hours. How long will it take her little sister to plant the flowers alone?

Step 1: We don't know how long it will take Mary's little sister to do the job alone. We assign x to represent this time. Then the little sister's rate is $\frac{1}{x}$.

Step 2: Mary can do the job in 3 hours, so we can state that her rate is $\frac{1}{3}$ of the job per hour. Together with her little sister, they can do the job in 2 hours, so we can state that their joint rate is $\frac{1}{2}$ of the job per hour.

Step 3: When we write the equation, the unknown is in the second term, since we don't know the time of the second person to finish the job:

$$\frac{1}{t_1}+\frac{1}{t_2}=\frac{1}{t_3}$$

Plug in the rates from Steps 1 and 2:

$$\frac{1}{3}+\frac{1}{x}=\frac{1}{2}$$

Step 4: Solve the equation by multiplying each term by the LCD, which is *6x*:

$$\frac{6x\cdot1}{3}+\frac{6x\cdot1}{x}=\frac{6x\cdot1}{2}$$

Reduce by the common factors to eliminate fractions:

$$\frac{\overset{2}{\cancel{6}}x}{\underset{1}{\cancel{3}}}+\frac{6\cancel{x}}{\cancel{x}}=\frac{\overset{3}{\cancel{6}}x}{\underset{1}{\cancel{2}}}$$

Multiply the remaining factors:

2x + 6 = *3x*

Isolate the variable:

2x – *3x* = –6

–*x* = –6

Divide by –1:

x = 6 hours

Solution: It will take the little sister 6 hours to plant the flowers alone.

Sometimes the time isn't given in exact hours. Instead, it is given in fractions of hours. We then need to use *complex fractions* to solve the problem.

DEFINITION

A **complex fraction** is a fraction that has a fraction in either the numerator or the denominator, or both, such as here:

$$\frac{\frac{2}{5}}{12},\ \frac{7}{\frac{3}{4}},\ \frac{\frac{5}{6}}{\frac{8}{9}}$$

Sample Problem 3

It takes one worker 3 hours to paint the fence. Another worker can paint the same fence in $2\frac{1}{4}$ hours. How long will it take to paint the fence if the two workers paint together?

The time for the second worker is given in hours and fractions of hours. Obviously, we could convert hours and fractions of hours into minutes for both workers and solve the equation to find time in minutes. Often this is not the best choice, though, since the equation would contain large denominators and a LCD that would be even larger. Sometimes it is easier to deal with complex fractions and have a more simple equation to solve.

Step 1: We don't know how long it will take to paint the fence if the two workers paint together, so we just say that this time is x. Then the rate of the two workers working together will be $\frac{1}{x}$. Since the first worker alone can do the job in 3 hours, his rate is $\frac{1}{3}$.

Step 2: Before we calculate the rate for the second worker, let's convert his time into an improper fraction: $2\frac{1}{4} = \frac{9}{4}$ hours. Since it takes him $\frac{9}{4}$ hours to finish the job alone, his rate is:

$$\frac{1}{\frac{9}{4}} = 1 \div \frac{9}{4} = \frac{1}{1} \cdot \frac{4}{9} = \frac{4}{9}$$

We found that the rate of the second worker is $\frac{4}{9}$.

Note that the whole job is still 1; we have 4 in the numerator because of the complex fraction we are dealing with.

Step 3: We know all the rates, so we can set up the equation:

$$\frac{1}{t_1} + \frac{1}{t_2} = \frac{1}{t_3}$$

Plug in the rates from Steps 1 and 2:

$$\frac{1}{3} + \frac{4}{9} = \frac{1}{x}$$

As we predicted, using complex fractions helped to obtain an easy equation with small numbers as the denominators and a very easy LCD, which is $9x$.

Step 4: To solve the equation, multiply each term by the LCD, which is $9x$:

$$\frac{9x \cdot 1}{3} + \frac{9x \cdot 4}{9} = \frac{9x \cdot 1}{x}$$

Reduce by common factors to eliminate fractions:

$$\frac{\overset{3}{\cancel{9}}x}{\underset{1}{\cancel{3}}} + \frac{\cancel{9}x \cdot 4}{\cancel{9}} = \frac{9\cancel{x}}{\cancel{x}}$$

Multiply the remaining factors and collect like terms:

$3x + 4x = 9$

$7x = 9$

Divide each side by 7:

$$x = \frac{9}{7} = 1\frac{2}{7} \text{ hours}$$

Solution: It will take $1\frac{2}{7}$ hours for the two workers to paint the fence together.

A WORD OF ADVICE

Simplifying a complex fraction means rearranging it into an equivalent simple fraction by dividing the numerator by the denominator:

$$\frac{\frac{5}{6}}{\frac{8}{9}} = \frac{5}{6} \div \frac{8}{9} = \frac{5}{\underset{2}{\cancel{6}}} \cdot \frac{\overset{3}{\cancel{9}}}{8} = \frac{15}{16}$$

Sample Problem 4

It takes Jim $\frac{4}{5}$ as long as it takes Bob to remodel the new customers' house. Working together, they can do the job in 200 hours. How long will it take each alone to do the job?

Step 1: Since we don't know either Jim's or Bob's time alone, let's assign x to represent Bob's time. Then Bob's rate will be $\frac{1}{x}$.

Step 2: We know that Jim's time is $\frac{4}{5}$ of Bob's time (x), so we need to multiply $\frac{4}{5}$ by x (remember that *of* means multiplication in algebra). We found that Jim's time to finish the job is $\frac{4}{5} \cdot x = \frac{4x}{5}$. To find his rate, we need to divide the whole job (1) by his time:

$$\frac{1}{\frac{4x}{5}} = 1 \div \frac{4x}{5} = \frac{1}{1} \cdot \frac{5}{4x} = \frac{5}{4x}$$

We found that Jim's rate is $\frac{5}{4x}$.

Step 3: Finally, we can express the rate when both are working together. Since together they can do the job in 200 hours, the rate is $\frac{1}{200}$.

Step 4: We know all the rates and can write the equation:

$$\frac{1}{t_1} + \frac{1}{t_2} = \frac{1}{t_3}$$

Plug in the rates from Steps 1, 2, and 3:

$$\frac{5}{4x} + \frac{1}{x} = \frac{1}{200}$$

To solve the equation, multiply each term by the LCD, which is $200x$:

$$\frac{200x \cdot 5}{4x} + \frac{200x \cdot 1}{x} = \frac{200x \cdot 1}{200}$$

Reduce by the common factors to eliminate fractions:

$$\frac{\overset{50}{\cancel{200}} \cancel{x} \cdot 5}{\cancel{4} \cdot \cancel{x}} + \frac{200 \cdot \cancel{x}}{\cancel{x}} = \frac{\cancel{200} \cdot x}{\cancel{200}}$$

Multiply the remaining factors:

$250 + 200 = x$

Collect like terms:

$x = 450$ hours

Since we denote by x Bob's time, we found that Bob alone can do the job in 450 hours.

Step 5: To find the time for Jim, we multiply Bob's time by $\frac{4}{5}$, since his time is $\frac{4}{5}$ of Bob's time:

$$\frac{4}{5}x = \frac{4}{5} \cdot 450 = \frac{4 \cdot \overset{90}{\cancel{450}}}{\underset{1}{\cancel{5}}} = 360 \text{ hours}$$

We found that Jim alone can finish the job in 360 hours.

Solution: Jim can finish the job in 360 hours alone, and Bob can finish the job in 450 hours alone.

In some problems, two people work together for some amount of time, then one leaves, and the other needs to finish the job alone; we need to find the time it will take to finish the job. These problems are a little bit trickier, but the routine is the same: find the rates or amount of the job per unit of time.

DEAD ENDS

In work problems in which "one person does part of the job and leaves," trying to set up the equation using the main work formula right away doesn't help. We need to find a portion of the job done before the person leaves and a portion of the job after the person leaves.

Sample Problem 5

Two mechanics are working to repair machinery. The more experienced mechanic can complete the job alone in 8 hours. The other one takes 12 hours to finish the job alone. They work together for the first 2 hours, and then the more experienced mechanic leaves to teach new guys. How long will it take for the other worker to finish the job?

Step 1: Let x indicate the time that the other worker needs to finish the job after the first worker left (this is what the problem wants us to find). Before we come back to discuss this portion of the work, let's deal with the portion of the job done by two workers before the more experienced one left.

Step 2: We know that the experienced worker does the whole job in 8 hours, so his rate is $\frac{1}{8}$. The other worker can do the whole job in 12 hours, so his rate is $\frac{1}{12}$.

Step 3: Together, their combined rate is:

$$\frac{1}{8} + \frac{1}{12} = \frac{3}{24} + \frac{2}{24} = \frac{5}{24}$$

Here, 24 is the LCD.

Since the rate is the amount of work done per hour, both workers can do $\frac{5}{24}$ of the job in 1 hour.

Step 4: They worked together for 2 hours, so they will do two times more work in 2 hours. We need to multiply their rate, which is $\frac{5}{24}$, by 2 hours:

$$2 \cdot \left(\frac{5}{24}\right) = \frac{\cancel{2}^{1} \cdot 5}{\cancel{24}_{12}} = \frac{5}{12}$$

We found that, in 2 hours, both workers got $\dfrac{5}{12}$ of the job done.

Step 5: Since the whole job is always 1, in order to find how much work is left, we need to subtract the portion of the job done by the two workers in 2 hours ($\dfrac{5}{12}$) from 1:

$$1 - \frac{5}{12} = \frac{12}{12} - \frac{5}{12} = \frac{7}{12}$$

We found that the other worker still needs to do $\dfrac{7}{12}$ of the job alone after the more experienced worker leaves.

Step 6: We know that, by x, we indicated the number of hours the second worker needs to finish this $\dfrac{7}{12}$ of the job. We can set up the following equation using the second worker's rate from Step 2:

Job per hour (rate) · Hours = Portion of the job done

$$\frac{1}{12} \cdot x = \frac{7}{12}$$

Step 7: Solve the equation to find the time needed to finish the job:

$$\frac{1}{12}x = \frac{7}{12}$$

$$\frac{x}{12} = \frac{7}{12}$$

Since the two fractions are equal and have the same denominator (12), they must have the same numerator as well.

$x = 7$

We found that the other worker has to work an additional 7 hours to finish the job.

Solution: The other mechanic will finish the job in 7 hours.

More Than Two People or Objects

More than two people can perform the same job simultaneously. The more people involved, the faster they can accomplish the job. For example, a team of three workers can mow the lawn much quicker than one or even two of them. The main work formula can be extended for more than two persons working together:

$$\frac{1}{t_1} + \frac{1}{t_2} + \frac{1}{t_3} = \frac{1}{t_4}$$

t_1 is time taken by the first person.

t_2 is time taken by the second person.

t_3 is time taken by the third person.

t_4 is time taken by the three of them working together.

WORTHY TO KNOW

The work formula can be extended for pipes and machinery as well and can include three or more objects working at the same time.

Sample Problem 6

Three people can finish waxing the floor in a building in 4 hours. If Larry does the job alone, he can finish the job in 10 hours. If Christopher does the job alone, he can finish it in 12 hours. How long will it take for the third worker, Mark, to finish the job alone?

Step 1: We don't know how long it will take Mark to finish the job, so let x indicate this time. Then his rate is $\dfrac{1}{x}$.

Step 2: Larry can finish the job in 10 hours, so his rate is $\dfrac{1}{10}$. Christopher can finish the job in 12 hours, so his rate is $\dfrac{1}{12}$. All three can finish the job in 4 hours, so their rate is $\dfrac{1}{4}$.

Step 3: We know all the rates, and we can set up an equation using the main work formula for three people:

$$\frac{1}{t_1} + \frac{1}{t_2} + \frac{1}{t_3} = \frac{1}{t_4}$$

Plug in the rates from Steps 1 and 2:

$$\frac{1}{10} + \frac{1}{12} + \frac{1}{x} = \frac{1}{4}$$

Step 4: To solve, multiply each term by the LCD, which is $60x$.

$$\frac{60x \cdot 1}{10} + \frac{60x \cdot 1}{12} + \frac{60x \cdot 1}{x} = \frac{60x \cdot 1}{4}$$

Reduce by common factors to eliminate fractions:

$$\frac{\overset{6}{\cancel{60}}x}{\cancel{10}} + \frac{\overset{5}{\cancel{60}}x}{\cancel{12}} + \frac{60\cancel{x}}{\cancel{x}} = \frac{\overset{15}{\cancel{60}}x}{\cancel{4}}$$

Multiply the remaining factors:

$6x + 5x + 60 = 15x$

Isolate the variable and collect like terms:

$6x + 5x - 15x = -60$

$-4x = -60$

Divide both sides by -4:

$x = 15$ hours

Solution: It will take Mark 15 hours to finish the job alone.

Pipes Filling and Draining Tanks and Pools

Another variation of work problems is filling and draining tanks and pools. When you have two or more pipes, the job of filling tanks and pools can be done faster unless one of the pipes is working against the other pipes and drains instead of filling. Then the rate of that pipe should be subtracted from, not added to, the rate of the other pipe(s).

DEAD ENDS

Because subtraction is not commutative, be careful to get the right order of the terms in the main work formula when you need to subtract.

Sample Problem 7

A tank can be filled by one pipe in 30 hours and emptied by another pipe in 40 hours. How long will it take to fill the tank if the two pipes are working at the same time?

Step 1: Since we don't know the time to fill the tank when both pipes are open, let x represent the time taken to fill the tank.

Step 2: The first pipe can fill the tank in 30 hours, so its rate is $\frac{1}{30}$ of the job in 1 hour. The second pipe empties the tank in 40 hours, so its rate is $\frac{1}{40}$ of the job per hour.

Step 3: The job is to fill the tank. Only the first pipe is doing this job. The second pipe is not efficient, since it works against the first pipe (emptying vs. filling). To indicate this fact, the rate of the second pipe must be subtracted from the rate of the first pipe in the main work equation:

$$\frac{1}{t_1} - \frac{1}{t_2} = \frac{1}{x}$$

Step 4: Substitute the rates from Step 2:

$$\frac{1}{30} - \frac{1}{40} = \frac{1}{x}$$

Step 5: To solve, multiply each term by the LCD, which is $120x$:

$$\frac{120x \cdot 1}{30} - \frac{120x \cdot 1}{40} = \frac{120x \cdot 1}{x}$$

Reduce to eliminate fractions:

$$\frac{\overset{4}{\cancel{120}}x}{\underset{1}{\cancel{30}}} - \frac{\overset{3}{\cancel{120}}x}{\underset{1}{\cancel{40}}} = \frac{120 \cdot \cancel{x}}{\cancel{x}}$$

Multiply the remaining factors:

$4x - 3x = 120$

Collect like terms on the left side:

$x = 120$ hours

Solution: It will take 120 hours to fill the tank if the two pipes are open at the same time.

Sample Problem 8

A tank can be filled by one pipe in 25 minutes and emptied by another pipe in 20 minutes. How long will it take to empty the tank if the two pipes are open at the same time?

Step 1: The job in this problem is to empty the tank. Since we don't know the time needed to do this job when the two pipes are open, let's assign x to represent this time. Then the rate will be $\dfrac{1}{x}$ of the job per minute.

Step 2: The first pipe can fill the tank in 25 minutes, so its rate is $\dfrac{1}{25}$ of the job per minute. The second pipe can empty the tank in 20 minutes, so its rate is $\dfrac{1}{20}$ of the job per minute.

Step 3: Since the job is to empty the tank, only the second pipe is doing the real job, while the first pipe is working against the second pipe and filling up the tank. That means the rate of the first pipe should be subtracted from the rate of the second pipe in the main work formula:

$$\frac{1}{t_2} - \frac{1}{t_1} = \frac{1}{t_3}$$

A WORD OF ADVICE

Always subtract the rate of a person or object working against the job from the rate of a person or object working toward finishing the job.

Step 4: Substitute the rates from Step 2:

$$\frac{1}{20} - \frac{1}{25} = \frac{1}{x}$$

To solve, multiply each term by the LCD, which is $100x$:

$$\frac{100x \cdot 1}{20} - \frac{100x \cdot 1}{25} = \frac{100x \cdot 1}{x}$$

Reduce to eliminate fractions:

$$\frac{\overset{5}{\cancel{100}}x}{\underset{1}{\cancel{20}}} - \frac{\overset{4}{\cancel{100}}x}{\underset{1}{\cancel{25}}} = \frac{100\cancel{x}}{\cancel{x}}$$

Multiply the remaining factors:

$5x - 4x = 100$

Collect like terms on the left side:

$x = 100$ minutes

Solution: It will take 100 minutes to empty the tank if the two pipes are open at the same time.

Sample Problem 9

At the chemical factory, a tank of acid can be filled by pipe A in 6 hours and by pipe B in 10 hours. The full tank of acid can be drained by pipe C in 8 hours. If the tank is empty and all three pipes are working at the same time, how long will it take to fill up the tank?

In this problem, the first two pipes are doing the job of filling the tank, while the third pipe is working against them.

Step 1: We don't know the time it takes to fill the tank with three pipes working, so let x indicate this time. If x is the time to fill the tank when all three pipes are working, then the combined rate for all three pipes is $\frac{1}{x}$ of the job in 1 hour.

Step 2: If pipe A can fill the tank in 6 hours, then its rate is $\frac{1}{6}$ of the job per hour. Pipe B can fill the tank in 10 hours, so its rate is $\frac{1}{10}$ of the job per hour. Both pipe A and pipe B are working on filling the tank, so their rates will be added together.

Step 3: Pipe C is draining the tank, so it is working against pipe A and pipe B's efforts. Since pipe C can drain the tank in 8 hours, its rate is $\frac{1}{8}$ of the job (or, better, non-job) per hour. This rate must be subtracted from the sum of the rates of pipe A and pipe B.

Step 4: Now that we know all the rates, we can set up an equation in which we add the rates of the first two pipes and subtract the rate of the third pipe:

$$\frac{1}{t_1} + \frac{1}{t_2} - \frac{1}{t_3} = \frac{1}{t_4}$$

We can plug in the rates from Steps 1, 2, and 3:

$$\frac{1}{6} + \frac{1}{10} - \frac{1}{8} = \frac{1}{x}$$

Step 5: To solve the equation, multiply each term by the LCD, which is $120x$:

$$\frac{120x \cdot 1}{6} + \frac{120x \cdot 1}{10} - \frac{120x}{8} = \frac{120x}{x}$$

Reduce by the common factors to eliminate fractions:

$$\frac{\overset{20}{\cancel{120}x}}{\underset{1}{\cancel{6}}} + \frac{\overset{12}{\cancel{120}x}}{\underset{1}{\cancel{10}}} - \frac{\overset{15}{\cancel{120}x}}{\underset{1}{\cancel{8}}} = \frac{120\cancel{x}}{\cancel{x}}$$

Multiply the remaining factors:

$20x + 12x - 15x = 120$

Collect like terms on the left side:

$17x = 120$

Divide both sides by 17:

$x = \dfrac{120}{17} = 7\dfrac{1}{17}$ hours

Solution: When all three pipes are working at the same time, it will take $7\dfrac{1}{17}$ hours to fill up the tank.

Practice Problems

Problem 1: One secretary can type a document in 30 minutes, while the other can type the same document in 45 minutes. How long will it take to type the document if the two of them are working together?

Problem 2: Two people work on a project. One of them, who is more experienced, will finish the project in 16 hours working alone. Both together will finish the project in 12 hours. How long will it take the second person, who is a new guy, to finish the job if he is working alone?

Problem 3: If a mechanic can repair a car in 12 hours, how much of the job is done after 10 hours?

Problem 4: It takes Anna twice as long to clean the kitchen as it does her older sister, Amy. How long will it take each girl by herself if they can clean the kitchen together in 40 minutes?

Problem 5: Fred can repair his car in $4\dfrac{1}{2}$ hours. His brother-in-law can do the same job in 4 hours. Fred's nephew, who is a professional mechanic, can finish the job in $3\dfrac{1}{2}$ hours. How long will it take to repair the car if all three work together?

Problem 6: An inlet pipe can fill a tank in 24 minutes. A drain pipe can drain the tank in 20 minutes. If the tank is full and both pipes are open, how long will it take until the tank is empty again?

Problem 7: One pipe can fill a tank in 12 minutes, and another pipe can fill it in 8 minutes. The third pipe can empty the tank in 10 minutes. With all pipes open, how long will it take to fill the tank?

The Least You Need to Know

- The main formula for work problems is $\frac{1}{t_1} + \frac{1}{t_2} = \frac{1}{t_3}$.
- The main formula for work can be expanded to include more people or objects doing the job.
- The job is always 100% = 1.
- Rate is the amount of the job done per unit of time. When a person can do a job in 4 hours, then he can do $\frac{1}{4}$ of the job in one hour (the whole job is 1). This means that his rate is $\frac{1}{4}$ of the job per hour.

Money, Mixtures, and Discount Problems

A great deal of our lives revolves around money: we go to work to earn it, we invest money to multiply its amount, and we go shopping to spend it. We can't do anything without money in modern times, so we need to learn how to deal with money: count it properly, invest with a good profit, and spend it wisely.

In this part, we discuss problems that one way or another relate to money. First we learn how to calculate the amount of special types of coins or bills using one or two equations. While discussing coins, we tackle problems about dry mixtures because you can use the same solving techniques for them as for the coin problems.

Later in this part, we explore problems that deal with finances: earning money with simple or compound interest and investing with profit or loss. Finally, we work through discount problems and discuss mathematical concepts and formulas that help calculate percent of discount, the original price, and the sale price.

Coins and Mixtures

In This Chapter

- Working with the number and value of coins
- Finding the total amount of money
- Using two variables to count coins
- Using a penny or dollar system to represent the value of each coin
- Dealing with dry mixtures

We deal with money every day of our lives—we earn it, spend it, or lose it. From the earliest days of civilization, the fact that we have money in our pockets makes us feel secure. But how much money do we have? Each type of coin or bill needs to be counted separately since they have different denominations. In this chapter, we discuss how to count money using one equation or a system of two equations. Later in the chapter, we explore dry mixtures problems since their solutions are similar to ones for coin problems.

Coins and One Equation

Counting coins involves more than just knowing that you have 20 coins in your pocket. Coins have different monetary values, so each type of coin needs to be counted separately. To calculate the total amount of money you have in coins or bills, you multiply the number of each by their monetary value and then add all the products to get the total amount of money.

DEAD ENDS

Don't look for *x* or *y* as variables in problems related to coins in this chapter. For convenience reasons, we use the letters *d, n,* and *q* to represent the number of dimes, nickels, and quarters.

Sample Problem 1

Bob has 9 more nickels than dimes. If the value is $3.90, how many of each coin does he have?

Step 1: Let the number of dimes be *d*. Since Bob has 9 more nickels, the number of nickels is *d* + 9.

Step 2: Convert dollars into cents:

$3.90 = 390¢

Step 3: Set up a table with quantity and value, filling in the Quantity column with information from Step 1. Put the values for each type of coin in the Value column. To obtain the Total column expression, multiply the quantity by the value for each type of coin:

	Quantity	Value (cents)	Total
Nickels	*d* + 9	5	5(*d* + 9)
Dimes	*d*	10	10*d*
Together			390

Step 4: We add the total for nickels with the total for dimes to get the total together:

5(*d* + 9) + 10*d* = 390

Distribute the left side and collect like terms:

5*d* + 45 + 10*d* = 390

15*d* = 390 − 45

15*d* = 345

Divide both sides by 15:

d = 23

We found that Bob has 23 dimes.

Step 5: To find the number of nickels, substitute the number of dimes (23) into the expression for nickels from Step 1:

$d + 9 = 23 + 9 = 32$

We found that Bob has 32 nickels.

Solution: Bob has 23 dimes and 32 nickels.

A WORD OF ADVICE

Setting up a table with quantity, value, and a total helps organize many coin problems and makes writing the equations easier.

Sample Problem 2

John received change worth $4. He received 2 more nickels than dimes and 10 fewer quarters than dimes. How many of each coin did he receive?

Step 1: Let variable d represent the number of dimes.

Since John has 2 more nickels than dimes, the number of nickels is $d + 2$. He also has 10 fewer quarters than the number of dimes, so the number of quarters can be expressed as $d - 10$. Also, we know that $4 = 400¢$.

Step 2: Set up a table with quantity and value, and fill in the Quantity column with information from Step 1. Put the values for each type of coin in the Value column. To get the Total column expression, multiply the quantity by the value for each type of coin:

	Quantity	Value (cents)	Total
Nickels	$d + 2$	5	$5(d + 2)$
Dimes	d	10	$10d$
Quarters	$d - 10$	25	$25(d - 10)$
Together			400

Step 3: We add the Total column to get the equation:

$5(d + 2) + 10d + 25(d - 10) = 400$

Distribute the left side and collect like terms:

$5d + 10 + 10d + 25d - 250 = 400$

$40d - 240 = 400$

$40d = 640$

Divide both sides by 40:

$d = 16$

We found that John has 16 dimes.

Step 4: To find the number of nickels, plug 16 into the expression for nickels from Step 1:

$d + 2 = 16 + 2 = 18$

To find the number of quarters, plug 16 into the expression for quarters from Step 1:

$d - 10 = 16 - 10 = 6$

We found that John has 18 nickels and 6 quarters.

Solution: John has 18 nickels, 16 dimes, and 6 quarters.

Sample Problem 3

Maria has 4 times as many $1 bills as she does $5 bills. She has a total of $54. How many of each bill does she have?

Step 1: Let the capital letter D represent the number of $5 bills. Since Maria has 4 times as many $1 bills, this number can be expressed as $4D$.

Step 2: Set up a table with quantity and value, and fill in the Quantity column with information from Step 1. Put the values for each type of bill in the Value column. For this problem, we use a dollar system to represent the value. To get the Total column expression, multiply the quantity by the value for each type of bill:

	Quantity	Value (dollars)	Total (dollars)
$1 bills	$4D$	1	$4D$
$5 bills	D	5	$5D$
Together			54

Step 3: We add the Total column to get the equation:

$4D + 5D = 54$

Collect like terms and divide both sides by 9:

$9D = 54$

$D = 6$

We found that Maria has 6 $5 bills.

DEAD END

Be consistent with units. For example, you might choose to represent the value of each coin based on a dollar system. Then the value would be in decimals—for example, the value of a nickel would be represented as 0.05. Alternately, you might choose a penny system. Then a nickel would be expressed as 5¢.

Step 4: To find the number of $1 bills, we need to multiply the number of $5 bills by 4, since Maria has 4 times more $1 bills:

$4D = 4 \cdot 6 = 24$

We found that Maria has 24 $1 bills.

Solution: Maria has 24 $1 bills and 6 $5 bills.

Coins and Two Equations

With this type of coin problem, you are given two statements. One statement is about the number of coins or bills. You need to convert it into an equation. The second statement is about the value of the coins and bills. You need to convert this statement into an equation about the values of the coins or bills that states the values in the same unit (penny system or dollar system). You then need to solve a system of two equations.

Sample Problem 4

Cindy has 41 coins consisting of nickels and dimes. The total value of the coins is $3.20. How many coins of each kind does she have?

Step 1: We don't know the number of either nickels or dimes. Let the number of nickels be n and the number of dimes be d. The total amount of money is $3.20—or, in the penny system, 320¢.

Step 2: Set up a table with quantity and value, and fill in the information from Step 1. Put the values for each type of coin in the Value column. To get the Total column expression, multiply the quantity by the value for each type of coin:

	Quantity	Value (cents)	Total
Nickels	n	5	$5n$
Dimes	d	10	$10d$
Together	41		320

Step 3: Add the Quantity column together to get the first equation. This equation expresses mathematically that Cindy has 41 coins (n nickels and d dimes):

$n + d = 41$

The equation has two unknowns and can't be solved alone. We need to combine it with another equation that has the same unknowns.

WORTHY TO KNOW

One of the oldest coins is called Lydian Trite, made of a gold and silver alloy. It was minted in Lydia, Asia Minor (present-day Turkey), around 600 B.C.E.

Step 4: Add up the Total column to get the second equation:

$5n + 10d = 320$

This equation expresses the statement about the amount of money Cindy has.

Step 5: Combine the equations from Steps 3 and 4 to get the system of linear equations:

$n + d = 41$

$5n + 10d = 320$

Step 6: We solve the system of two equations by substitution method. Revise the first equation:

$n = 41 - d$

Step 7: Substitute the revised first equation into the second equation for n:

$5(41 - d) + 10d = 320$

Distribute the left side and collect like terms:

$205 - 5d + 10d = 320$

$205 + 5d = 320$

Isolate the variable:

$5d = 115$

Divide both sides by 5:

$d = 23$

We found that Cindy has 23 dimes.

Step 8: To find the number of nickels, substitute the value of d (23) into the revised equation in Step 6:

$n = 41 - d = 41 - 23 = 18$

We found that Cindy has 18 nickels.

Solution: Cindy has 18 nickels and 23 dimes.

A WORD OF ADVICE

It is always advisable to plug your answers back into the problem to check that your solution is correct.

Sample Problem 5

Michael had $12.45 in quarters and dimes. If he had 87 coins in all, how many of each type of coin did he have?

Step 1: We don't know the number of either quarters or dimes. Let the number of quarters be q and the number of dimes be d. The total amount of money is $12.45—or, in the penny system, 1,245¢.

Step 2: Set up a table with quantity and value, and fill in the information from Step 1. Put the values for each type of coin in the Value column. To get the Total column expression, multiply the quantity by the value for each type of coin:

	Quantity	Value (cents)	Total
Quarters	q	25	$25q$
Dimes	d	10	$10d$
Together	87		1,245

Step 3: Add the Quantity column to get the first equation. This equation expresses mathematically that Michael had 87 coins:

$q + d = 87$

The equation has two unknowns and can't be solved alone. We need another equation to set up a system of two equations.

Step 4: Add up column 4 to get the second equation:

$25q + 10d = 1,245$

This equation expresses the statement about the amount of money Michael had.

Step 5: Combine the equations from Steps 3 and 4 to get the system of linear equations:

$q + d = 87$

$25q + 10d = 1,245$

Step 6: We solve the system of two equations using the substitution method. Revise the first equation:

$q = 87 - d$

Step 7: Substitute the revised first equation into the second equation for q:

$25(87 - d) + 10d = 1,245$

Distribute the left side and collect like terms:

$2,175 - 25d + 10d = 1,245$

$2,175 - 15d = 1,245$

Isolate the variable:

$-15d = -930$

Divide both sides by -15:

$d = 62$

We found that Michael had 62 dimes.

Step 8: To find the number of quarters, substitute the value of d (62) into the revised equation in Step 6:

$q = 87 - d = 87 - 62 = 25$

We found that Michael had 25 quarters.

Solution: Michael had 62 dimes and 25 quarters.

WORTHY TO KNOW

Similar to coin problems are stamp problems and ticket problems. They can be solved using the same technique.

Sample Problem 6

Tickets to the community theater cost $15 for adults and $7 for children. For last week's performance, a total of 310 tickets were sold for $4,210. How many tickets of each kind were sold?

Step 1: We don't know the number of child or adult tickets sold. Let the number of child tickets sold be x, and the number of adult tickets sold be y. The total amount of money gained after selling all tickets is $4,210.

Step 2: Set up a table with quantity and price, and fill in the information from Step 1. Put the price in dollars for each type of ticket in the Price column. To get the total column expression, multiply the quantity by the price for each type of ticket:

	Quantity	Price (dollars)	Total
Children	x	7	$7x$
Adults	y	15	$15y$
Together	310		4,210

Step 3: Combine the Quantity column to get the first equation. This equation expresses mathematically that 310 tickets were sold:

$x + y = 310$

The equation has two unknowns and can't be solved alone. We must get another equation to set up a system of two equations.

Step 4: Add the Total column to get the second equation:

$7x + 15y = 4,210$

This equation expresses the statement about the amount of money for selling 310 tickets.

Step 5: Combine the equations from Steps 3 and 4 to get the system of linear equations:

$x + y = 310$

$7x + 15y = 4{,}210$

Step 6: We can solve the system of two equations using the substitution method. Revise the first equation:

$y = 310 - x$

Step 7: Substitute the revised first equation into the second equation for y:

$7x + 15(310 - x) = 4{,}210$

Distribute the left side and collect like terms:

$7x + 4{,}650 - 15x = 4{,}210$

$-8x = -440$

Divide both sides by -8:

$x = 55$

We found that 55 child tickets were sold.

Step 8: To find the number of adult tickets sold, substitute the value of $x(55)$ into the revised equation in Step 6:

$y = 310 - x = 310 - 55 = 255$

We found that 255 adult tickets were sold.

Solution: We found that 55 child tickets and 255 adult tickets were sold.

Mixing Quantities of Different Price

There are two different types of mixtures: dry mixtures and liquid mixtures. We deal with liquid mixtures in Chapter 16. For now, we concentrate on dry mixtures. Dry mixtures contain two differently priced components, such as two types of nuts or candies. To solve dry mixture problems, we need to set up an equation showing that the total cost of the mixture is the sum of the total cost of the first component plus the total cost of the second component.

Sample Problem 7

A manager of a candy store mixes two types of candies. He decides to create a 40-pound mixture of jelly beans and Milk Duds. The jelly beans sell for $1.90 per pound, and the Milk Duds sell for $2.40 per pound. He plans to sell the mix for $2.20 per pound. How many pounds of each candy should he use in the mix?

Step 1: The manager knows that he wants the total weight of the mixture to be 40 pounds. However, he doesn't know how many pounds of each candy type to mix. This is the problem's question. We use x to represent the weight of the jelly beans. The weight of the Milk Duds can be presented as $40 - x$ (total pounds of candy minus x).

 A WORD OF ADVICE

Using a table for mixture problems is a helpful way to illustrate the problem.

Step 2: Set up a table with four columns and four rows:

Candy Type	Cost of Candy (unit price)	Amount of Candies (in pounds)	Total Cost (in dollars)
Jelly beans	1.90	x	$1.90x$
Milk Duds	2.40	$40 - x$	$2.40(40 - x)$
Mixture	2.20	40	$(2.20)(40)$

The first column shows the type of candy, and the second column displays cost per pound in dollars. Note that, for this problem, we use a dollar system, so cost is given in dollars, not cents. The third column lists the weight of each type of candy found in Step 1 and given by the problem (40 pounds). Finally, the fourth column displays the total cost for each type of candy. The total cost is found by multiplying the cost per pound by the weight of each type.

Step 3: The Total Cost column is used to write the equation:

The total cost of jelly beans plus the total cost of Milk Duds equals the total cost of the mixture.

$1.90x + 2.40(40 - x) = (2.20)(40)$

Step 4: We are ready to solve the equation and find the weight of each type of candy in the mixture:

$1.90x + 2.40(40 - x) = (2.20)(40)$

Multiply each term by 100 to eliminate decimals:

$190x + 240(40 - x) = (220)(40)$

Distribute to remove parentheses:

$190x + 9,600 - 240x = 8,800$

Isolate the variable and collect like terms:

$-50x = -800$

Divide both sides by -50:

$x = 16$ pounds

We found that the manager used 16 pounds of jelly beans.

Step 5: To find the amount of Milk Duds, we use the expression from Step 1: $(40 - x)$. We substitute 16 for x:

$40 - x = 40 - 16 = 24$ pounds

We found that the manager used 24 pounds of Milk Duds.

Solution: The manager used 16 pounds of jelly beans and 24 pounds of Milk Duds.

WORTHY TO KNOW

Many problems about coins, mixtures, stamps, and tickets can be solved using either one equation or a system of two equations. This depends on the way we choose variables.

Sample Problem 8

How many pounds of peanuts for $2.60 per pound must be mixed with 12 pounds of pine nuts selling for $7.40 per pound to create a nut mixture that sells for $5.80 per pound?

Step 1: We know that the amount of pine nuts in the mixture is 12 pounds. We don't know how many pounds of peanuts to use. This is the problem's question. We can use x to represent the amount of peanuts. Then the amount of the nut mixture can be presented as $x + 12$.

Step 2: Set up a table with four columns and four rows:

Nut Type	Cost of Nuts (unit price)	Amount of Nuts (in pounds)	Total Cost (in dollars)
Peanuts	2.60	x	$2.60x$
Pine nuts	7.40	12	$(7.40)(12)$
Mixture	5.80	$x + 12$	$(5.80)(x + 12)$

The first column shows the type of nuts, and the second column displays cost per pound in dollars. We again are using the dollar system, so the cost is given in dollars. The third column lists the amount of each type of nut found in Step 1 and given by the problem (12 pounds). Finally, the fourth column displays the total cost for each type of nut. The total cost is found by multiplying the cost per pound by the amount of each type.

Step 3: The Total Cost column is used to write the equation:

The total cost of peanuts plus the total cost of pine nuts equals the total cost of the mixture.

$2.60x + (7.40)(12) = (5.80)(x + 12)$

Step 4: We are ready to solve the equation to find the amount of peanuts in the mixture:

$2.60x + (7.40)(12) = (5.80)(x + 12)$

Multiply each term by 100 to eliminate decimals:

$260x + 740(12) = (580)(x + 12)$

Distribute the left side to remove parentheses:

$260x + 8,880 = 580x + 6,960$

Isolate the variable and collect like terms:

$-320x = -1,920$

Divide both sides by -320:

$x = 6$ pounds

We found that we need to use 6 pounds of peanuts.

Solution: We need to use 6 pounds of peanuts.

Practice Problems

Problem 1: Tom has some amount of dimes and quarters totaling $8.30. The number of quarters is 6 more than 3 times the number of dimes. How many coins of each kind does he have?

Problem 2: A coin purse contains a mixture of nickels, dimes, and quarters. The coins have a total value of $4.25. There are 3 times more nickels than the number of dimes, and the number of quarters is 7 more than the number of dimes. Determine the number of nickels, dimes, and quarters in the purse.

Problem 3: Mary bought 30 stamps for $4.98. Some of them were 34¢ stamps and the rest were 5¢ stamps. How many of each kind did she buy?

Problem 4: Larry has $1.80 in change in his pocket consisting of nickels and dimes. If he has a total of 26 coins, how many of each kind does he have in his pocket?

Problem 5: A store that sells tea plans to mix a more expensive tea that costs $6 per pound with a less expensive tea that costs $4 per pound to create a 100-pound blend that will sell for $5 per pound. How many pounds of each type of tea are required?

Problem 6: How many pounds of white chocolate candies selling for $4.48 per pound must be mixed with 20 pounds of nuts selling for $8.96 per pound to create a mixture that sells for $7.28 per pound?

Problem 7: Tickets to the community theater cost $18 for adults and $8 for children. A total of 280 tickets were sold for $4,590. How many tickets of each kind were sold?

The Least You Need to Know

- To calculate the total amount of money, multiply the number of each type of coin by their monetary value and then add all the products to get the total amount of money.
- Creating a table for coin problems helps to organize the problem. The Total column in the table usually helps in writing the equation.
- When two statements about coins are given, express each as an equation. Usually, the first statement is about the number of coins or bills; for example, "Bob has 15 coins consisting of nickels and dimes." The second statement is about the value of the coins or bills; for example, "the total value of Bob's coins is $11.50."
- The total cost of a mixture is the sum of the total cost of the first component plus the total cost of the second component.

Manage Your Finances

Chapter 12

In This Chapter

- Computing the total amount of money with simple interest
- Comparing money earned with compound interest
- Finding how much to put into an account to meet a goal
- Calculating earnings and losses on investments

Understanding how to manage finances is the foundation to achieving long-term goals such as buying a house or a car. Understanding finances also helps people avoid making poor financial decisions, like spending too much money and suddenly becoming caught in a serious financial crunch. Understanding how to save and invest money puts you in charge of your life, helps you set goals, and enables you to grow your money wisely.

Gaining Money with Simple Interest

When we deposit or borrow money, there are two dollar amounts involved: *principal* and *interest*. For example, when you deposit $3,000 in a savings account, the $3,000 is the principal, and the bank then pays you interest for the use of the money (lately only a paltry sum, but better than nothing). The other side of interest is that when you borrow, for example, $3,000, you have to pay interest for the privilege of using the money until you pay it off fully (likely at a much higher interest rate).

> **DEFINITION**
>
> The amount of money that you borrow or deposit is called the **principal.** The dollar amount that you pay for borrowing money or that you get paid for lending money is called the **interest.**

The amount of interest depends on the principal and the interest rate, which is given as a percent (and varies from bank to bank). The amount of interest you pay or get paid also depends on the length of time for which the money is deposited or borrowed. In this section of the chapter, the rate is assumed to be annual (per year).

Simple interest involves interest calculated only on the principal amount. We use the following formula to find simple interest:

Interest = Principal · Rate · Time

$I = Prt$

The rate, r, is expressed as a decimal when calculating simple interest, and t is time in years.

Sample Problem 1

You deposit $4,000 in a savings account (certificate of deposit, or CD) at Sunrise Bank, which has a rate of 5%. Find the interest at the end of the first year.

Step 1: The principal P is $4,000, and the rate r is 5%. Time t is 1 year. Let's change the percent into the decimal form:

5% = 0.05

Step 2: To find the interest I at the end of the first year, we use the simple interest formula and plug in the known values from Step 1:

$I = Prt = (4,000)(0.05)(1) = \$200$

We found that, at the end of the first year, the interest earned is $200 and the total amount of money in the account is $4,200 ($4,000 + $200 = $4,200). You can withdraw $200, and you will still have the same $4,000 in your savings account.

Solution: The interest is $200.

Sample Problem 2

Anna wants to have an interest income of $1,500 a year. How much must she deposit for 1 year at 8%?

Step 1: We know the interest I, which is $1,500, and the rate r, which is 8%. The time t is 1 year. We need to find the principal P, which is the unknown.

Step 2: Change the rate into decimal form:

8% = 0.08

Step 3: Use the simple interest formula and plug in the known values from Steps 1 and 2:

$I = Prt$

$1,500 = P(0.08)(1)$

$1,500 = 0.08P$

Divide both sides by 0.08 to find P:

$$P = \frac{\$1,500}{0.08} = \$18,750$$

Solution: Anna must deposit $18,750.

Sample Problem 3

If John earns $250 interest in 30 months on money he invested at 5%, find the original principal.

Step 1: We know the interest, which is $250. Let P represent the principal that we need to find. Let's also convert 30 months into years:

30 months = 24 months + 6 months = 2 years + 0.5 years = 2.5 years

Step 2: Convert the interest rate into the decimal form:

5% = 0.05

A WORD OF ADVICE

When time in a problem is given in months, convert months into years and then substitute into the simple interest formula. For example, if the money is deposited for 18 months, you need to plug 1.5 years into the formula.

Step 3: We can find the principal P using the simple interest formula and substituting all the variables and values from Steps 1 and 2:

$I = Prt$

$250 = P(0.05)(2.5)$

$250 = P(0.125)$

To find P, divide both sides by 0.125:

$$P = \frac{250}{0,125} = \$2,000$$

Solution: The original principal was $2,000.

Sample Problem 4

If $600 is borrowed at an interest rate of 4% for 5 years, find the total amount of money to be paid off at the end of the fifth year.

Step 1: Let the capital letter A represent the total amount of money to be paid off. This amount of money involves two parts: the principal P ($600) and the interest I (unknown) accumulated over 5 years. We can express this statement mathematically as:

$A = P + I$

Step 2: We can find the interest I using the simple interest formula. Since the formula uses the percent in its decimal form, let's first change the interest rate into its decimal form:

4% = 0.04

Step 3: Plug the known values from Steps 1 and 2 into the simple interest formula to find the interest at the end of the fifth year:

$I = Prt = (600)(0.04)(5) = \$120$

Note that the time is 5 years, so we plug in 5 for the time t.

We found that the interest accumulated in 5 years is $120.

Step 4: Now we can calculate the total amount of money by substituting the value for interest from Step 3 and the known principal ($600) into the formula in Step 1:

$A = P + I = 600 + 120 = \$720$

Solution: The amount of money to be paid off is $720.

Sample Problem 5

The amount of annual interest earned with $30,000 is $150 less than the amount of annual interest earned with $36,000 at a 0.5% lower interest rate per year. What is the rate of interest on each amount of money?

Step 1: We don't know the interest rates, and this is the problem's question. We do know that the second rate is 0.5% lower than the first interest rate. Let's first change this difference into the decimal form, since we use the decimal form of the interest rates in the main formula:

$0.5\% = 0.005$

Let r_1 indicate the rate for the first principal; then the rate for the second principal can be expressed as $r_2 = (r_1 - 0.5\%) = (r_1 - 0.005)$, since it is 0.5% (0.005) lower than the first rate.

Step 2: We have two principals in this problem. The first principal is $P_1 = \$30,000$, and the second principal is $P_2 = \$36,000$. The time t is 1 year for both principals.

Step 3: We also know that the difference between the two interests is $150, since the first amount of money ($30,000) earned $150 less than the second amount of money ($36,000). We can express this fact mathematically as:

$I_2 - I_1 = \$150$

I_1 and I_2 are the interests on the first principal and the second principal, respectively.

Step 4: Now we can use the simple interest formula for both interests and plug in the variables and values from Steps 1 and 2:

$I_1 = P_1 r_1 t = (30,000)(r_1)(1) = 30,000 r_1$

$I_2 = P_2 r_2 t = (36,000)(r_1 - 0.005)(1) = 36,000(r_1 - 0.005)$

Step 5: Substitute the two expressions for interests from Step 4 into the formula in Step 3 to obtain the equation for the problem:

The interest on the first amount minus the interest on the second amount equals $150.

$36,000(r_1 - 0.005) - 30,000 r_1 = 150$

Distribute the left side:

$36,000 r_1 - 180 - 30,000 r_1 = 150$

Collect like terms on the left:

$6,000r_1 - 180 = 150$

Isolate the variable:

$6,000r_1 = 330$

Divide both sides by 6,000:

$r_1 = 0.055$

We found that the rate on the first principal was 0.055 in the decimal from.

Step 6: We need to change the decimal form into a percent:

$r = 0.055 = 5.5\%$

We found that the first interest rate r_1 was 5.5%.

To find the second interest rate, substitute r_1 (in percent form) into the expression for the second interest rate from Step 1:

$r_2 = (r_1 - 0.5\%) = (5.5\% - 0.5\%) = 5\%$

We found that the second interest rate was 5%.

Solution: The first interest rate was 5.5%, and the second interest rate was 5%.

Gaining or Losing Money with Compound Interest

The main difference between simple and compound interest is the fact that simple interest, as we have already mentioned, is paid only on the principal, whereas compound interest is paid on both the principal and the accumulated interest. Let's see the difference in earnings with the following example.

If you invest $3,000 at 5% interest for 2 years, you earn the following simple interest:

$I = Prt = (3,000)(0.05)(2) = \$300$

We found that you earned $300 in simple interest.

If you invest the same amount of money ($3,000) at the same rate (5%) and for the same period of time (2 years), this is how we can calculate compound interest:

At the end of the first year, the interest will be:

Interest year 1 = $(3,000)(0.05)(1) = \$150$

A WORD OF ADVICE

Compound interest is paid on both the principal and the accumulated interest.

Now we need to add this amount to the initial principal ($3,000) to find the new principal for the second year:

$3,000 + $150 = $3,150

Let's substitute the new principal into the formula. The interest at the end of the second year will be:

Interest year 2 = (3,150)(0.05)(1) = $157.50

The total compound interest at the end of the 2 years will be:

$150 + $157.50 = $307.50

This is $7.60 more than the amount of simple interest. It may not seem like a huge difference, but when big money is involved, the difference is very noticeable.

However, we don't need to calculate compound interest after each year separately and then add up the amounts for each year to find out the total interest. A formula helps us calculate the total amount of money:

$$A = P\left(1 + \frac{r}{n}\right)^{nt}$$

A is the amount of money after a certain period of time.

P is the principal, or the amount of money you start with.

r is the interest rate and must be in decimal form.

t is the amount of time in years.

n is the number of times interest is compounded in one year.

For example, $n = 1$ if interest is compounded annually; if interest is compounded quarterly, then $n = 4$; and if interest is compounded monthly, then $n = 12$.

The equation for compound interest involves five variables, four of which are given to you. Your job is to find the fifth variable.

Sample Problem 6

Carl deposited $1,000 at the interest rate of 4%. The interest is compounded quarterly, and the money will stay in the account for 5 years. How much money will be in Carl's account at the end of the fifth year?

Step 1: We need to find the amount of money A. We know that the principal P is $1,000. The number of years is 5, so $t = 5$. The interest is compounded quarterly, or four times per year, so $n = 4$. The rate is 4% and, in the decimal form, is 4% = 0.04.

Step 2: We use the formula for compound interest and substitute values from Step 1:

$$A = P\left(1 + \frac{r}{n}\right)^{nt}$$

$$A = 1{,}000\left(1 + \frac{0.04}{4}\right)^{(4)(5)}$$

$$A = 1{,}000(1 + 0.01)^{20}$$

$$A = 1{,}000(1.01)^{20} \approx 1{,}000(1.22019) \approx 1{,}220.19$$

We then round the answer to the hundredths place.

We found that, at the end of the fifth year, the account is worth $1,220.19 (that includes the principal and interest combined).

Solution: Carl will have $1,220.19 at the end of the fifth year.

Sample Problem 7

Mindy wanted to place $2,000 in a bank to earn interest over the next year. She found two banks that pay the same interest rate of 8%, but the first bank compounds monthly and the second bank pays only simple interest. What is the difference in the amount of interest she can earn at the two banks?

Step 1: Let's first deal with the bank that pays only simple interest. We know the principal amount is P = $2,000; the amount of time is 1 year, so t = 1. The rate is 8%, which, in decimal form, is: 8% = 0.08.

Step 2: We can use the formula for simple interest and substitute values from Step 1:

$I = Prt$

$I = (2,000)(0.08)(1) = \$160$

We found that Mindy can earn $160 at the bank that gives only simple interest.

Step 3: Let's now deal with the bank that compounds monthly. The formula for compound interest provides the total amount of money at the end of the year (balance), so we need to calculate this amount A first and then find the interest. The principal is still the same, P = $2,000; the rate is r = 8% = 0.08, time is t = 1, and n = 12 since the interest is compounded monthly.

Step 4: We can use the formula for compound interest and substitute the values from Step 3:

$$A = P\left(1 + \frac{r}{n}\right)^{nt}$$

$$A = 2,000\left(1 + \frac{0.08}{12}\right)^{(12)(1)}$$

$$A \approx 2,000\left(1 + 0.00667\right)^{12}$$

$$A \approx 2,000\left(1.00667\right)^{12} \approx 2.000\left(1.08304\right) \approx \$2,166.08$$

We rounded the answer to the hundredths place.

We found that, at the end of the first year, the total amount of money in the second bank's account would be $2,166.08.

A WORD OF ADVICE

Since we're dealing with money in these problems, it makes sense to always round to two decimal places and use the "approximately equal to" sign (≈) in our calculations.

Step 5: To find the interest earned at the bank that compounds interest monthly, we need to subtract the principal ($2,000) from the total amount of money we found in Step 4:

Interest at bank 2: $2,166.08 – $2,000 = $166.08

We found that the interest earned at the second bank is $166.08.

Step 6: Finally, we can answer the problem's question: what is the difference in the amount of interest Mindy can earn at the two banks? We need to subtract the interest earned at the bank that gives only simple interest from the interest earned from the bank that compounds interest monthly:

$166.08 – $160 = $6.08

Solution: The difference in the amount of interest is $6.08.

Did I Invest Properly?

Investment problems usually involve simple annual interest, so we use the simple interest formula $I = Prt$. In this formula, I stands for the interest on the original investment and P stands for the amount of the original investment. As always, r is the rate and t is the time.

When you invest money, especially with stocks, you agree to take some risk, since you either can gain profit or suffer a loss. In investment problems, profit and loss are expressed as percents, which you need to change into decimal form to calculate the amount of money gained or lost. To find the profit (in dollars) on an investment after 1 year, multiply the percent of profit r by the amount of the initial investment P:

Profit = Pr

When someone earns profits from two accounts, we add two profits to get the total amount of money earned.

A loss on an investment after 1 year can be calculated with the following formula:

Loss = Pr

Just remember to change percents into decimal form.

Sample Problem 8

Maria invested $5,000, part of it at 8% and the rest at 9%. How much did she invest at each rate if the total interest she earned for a year was $430?

We can solve this problem using a system of two equations.

Step 1: Let x represent the amount invested at 8%. Let y represent the amount invested at 9%. Since together she invested $5,000, we can express this fact mathematically as:

$x + y = 5,000$

This is our first equation; we need the second one, too, since we have two variables.

Step 2: To write the second equation, we use the information about the interest earned at both rates. First we express the first interest earned denoted as I_1 earned at the rate of 8%. Time is 1 year, the rate in the decimal form is 8% = 0.08, and the principal is x. We use the formula for simple interest and substitute these values:

$I_1 = Prt = x(0.08)(1) = 0.08x$

Step 3: Now we do the same with the second interest, which we denote as I_2. The principal is y, the time is 1 year, and the rate in decimal form is 9% = 0.09. We can plug these values into the formula:

$I_2 = Prt = y(0.09)(1) = 0.09y$

Step 4: We know that the total interest Maria earned from both investments was $430, so we can express this fact mathematically using the two expressions from Steps 2 and 3:

$I_1 + I_2 = 430$

$0.08x + 0.09y = 430$

This is our second equation. To simplify, multiply each term by 100:

$8x + 9y = 43,000$

Step 5: Combine both equations from Steps 1 and 4 to obtain a system of two equations:

$$\begin{cases} x + y = 5,000 \\ 8x + 9y = 43,000 \end{cases}$$

A WORD OF ADVICE

The problems on investments in this section and other similar problems can be solved using either one equation or a system of two equations. It depends on the way you choose variables.

Step 6: We can solve the system using the addition method.

Multiply both sides of the first equation by –8 and leave the second equation unchanged:

$$\begin{cases} -8x - 8y = -40,000 \\ 8x + 9y = 43,000 \\ \hline y = 3,000 \end{cases}$$

We found that the amount invested at the rate of 9% was $3,000.

Step 7: To find the amount invested at the rate of 8%, we use the expression from Step 1 and plug $3,000 into it:

$x + y = 5,000$

$x + 3,000 = 5,000$

$x = 5,000 - 3,000 = 2,000$

We found that the amount of money invested at the rate of 8% was $2,000.

Solution: Two thousand dollars was invested at the rate of 8% and $3,000 was invested at the rate of 9%.

Sample Problem 9

A woman made two investments totaling $15,000. On one investment she made a 12% profit, but the other part of money she invested in stock suffered a 10% loss. Find the amount of each investment if her overall net loss was $400.

We can solve this problem using one equation.

Step 1: The total amount of money the woman invested is $15,000. Let x represent the amount of money invested with a 12% profit. Then the amount of money invested in stock is $15,000 - x$.

DEAD ENDS

Don't add profit and loss to get the total result. If the net result is a profit, subtract the loss from the profit. If the net result is a loss, subtract the profit from the loss.

Step 2: We know that the woman made a 12% profit on her first investment. Twelve percent is 12% = 0.12 in decimal form. Using the formula for profit that we discussed at the beginning of this section and x as the principal for the first investment, we can calculate the amount of money the woman made in her first investment:

Profit = $Pr = x(0.12) = 0.12x$

Step 3: We can calculate the loss that the woman took on her second investment by substituting the expression $(15{,}000 - x)$ as the initial investment and 10% in the decimal form 10% = 0.1 into the formula:

Loss = $Pr = (15{,}000 - x)(0.1) = 0.1(15{,}000 - x)$

This is the amount of money the woman lost on her second investment.

Step 4: We know from the problem that even though the woman made money on her first investment, because of her loss on the second one, she took an overall net loss of $400. To write the equation, we need to translate the following statement into algebra language:

The amount lost minus the amount gained equals the net loss.

$0.1(15{,}000 - x) - 0.12x = 400$

Step 5: To solve the equation, first multiply both sides by 100:

$10(15{,}000 - x) - 12x = 40{,}000$

Distribute the left side:

$150{,}000 - 10x - 12x = 40{,}000$

Collect like terms:

$150{,}000 - 22x = 40{,}000$

Isolate the variable:

$-22x = -110{,}000$

Divide both sides by –22:

$x = 5{,}000$

We found that the woman invested $5,000 to her first investment.

Step 6: To find the amount of money in her second investment, we can use the expression from Step 1 and substitute 5,000 for x:

$15{,}000 - x = 15{,}000 - 5{,}000 = 10{,}000$

We found that the woman had $10,000 in stock.

Solution: The first investment was $5,000 and the second investment was $10,000.

Practice Problems

Problem 1: If Paul earns $300 interest in 42 months on money that was invested at 6%, find the original principal.

Problem 2: If $800 is borrowed at an interest rate of 5% for 4 years, find the total amount of money to be paid off at the end of the fourth year.

Problem 3: The amount of annual interest earned by $16,000 is $300 less than the amount of annual interest earned by $24,000 at 0.75% less interest rate per year. What is the rate of interest on each amount of money?

Problem 4: A man invested $5,000 in an account that pays 8.5% interest per year compounded quarterly. How much money will he have after 3 years?

Problem 5: Kelly wants to have a total of $8,000 in 2 years so she can remodel her kitchen. She finds an account that pays 5% interest compounded monthly. How much should Kelly put into this account so that she has $8,000 at the end of the second year?

Problem 6: A person plans to invest twice as much money into a First Mortgage account at a 3.5% annual interest as in a Second Mortgage account at an 8% annual interest. How much should the person invest in each account so that the total interest for one year will be $870?

Problem 7: George made two investments totaling $20,000. On one investment he made a 15% profit, but on the other he suffered a 12% loss. If his net gain was $840, how much was each investment?

The Least You Need to Know

- Simple interest involves interest calculated only on the principal amount.
- The main formula for simple interest is $I = Prt$.
- The rate of interest must be entered in decimal form into the formula to calculate interest.
- The main formula for compound interest is $A = P\left(1 + \dfrac{r}{n}\right)^{nt}$.
- The formulas to calculate investment profit or loss are: Profit = Pr and Loss = Pr.

Deep Discount

Chapter 13

In This Chapter

- Using two methods to solve discount problems
- The difference between wholesale and retail prices
- Understanding markup and markdown
- Dealing with decimals in equations

To entice us into the store and encourage us to buy goods, owners and managers of stores sell items on sale at a reduced price. When an item is placed on sale, the amount by which the price is reduced is called the *discount*. This chapter can help you to find all elements related to the concept of a discount: the original price, the sale price, and the percent of discount.

Finding Original Price

When an algebra word problem asks to find the original price when the percent of discount and discounted (sale) price is given, we can use one of the following methods:

DEFINITION

Discount is the amount by which the price is reduced. It is expressed as a percentage of the original price.

Method 1:

Step 1: Subtract the percent of discount from 100%:

100% – Percent of discount = Decreased percent

Step 2: Divide the result into the sale price to find the original price:

$$\frac{\text{sale price}}{\text{decreased percent}} = \text{original price}$$

Method 2:

Use this formula, where x stands for the original price you need to find and $p\%$ is the percent of discount in the decimal form:

$x - p\% \cdot x = $ Sale price

Solve the equation for x.

Obviously, using either method achieves the same answer.

Sample Problem 1

Sara paid $150 for a business suit. The sale price included a 25% discount. What was the original price of the suit?

We'll solve this problem using both methods.

Method 1:

Step 1: Let the original price be x.

Step 2: Subtract the percent of discount from 100% to find the decreased percent:

100% – 25% = 75%

Step 3: Change the decreased percent into decimal form:

75% = 0.75

Step 4: Divide the decreased percent into the sale price:

$$\frac{\text{sale price}}{\text{decreased percent}} = \text{original price}$$

$$\frac{150}{0.75} = \frac{15,000}{75} = \$200$$

Solution: The original price was $200.

Method 2:

Step 1: Let the original price be x. Let's also change the percent into decimal form:

25% = 0.25

Step 2: Use the formula:

$x - p\% \cdot x$ = Sale price

Substitute the value for the sale price ($150) and the discount percent in decimal form into the formula:

$x - 0.25x = \$150$

Step 3: Multiply each term by 100 to eliminate decimals:

$100x - 25x = \$15,000$

Collect like terms:

$75x = \$15,000$

Divide both sides by 75:

$x = \$200$

Solution: The original price was $200.

As we expected, both methods provided the same answers.

Sample Problem 2

An item is marked down 15%; the sale price is $127.50. What was the original price?

We'll solve this problem using the first method:

Step 1: Let the original price be x.

Step 2: Subtract the markdown rate from 100% to find the decreased percent:

100% − 15% = 85%

WORTHY TO KNOW

Markdown rate is the same as discount rate. Both are given in percents.

Step 3: Change the decreased percent to decimal form:

85% = 0.85

Step 4: Divide the decreased percent into the sale price:

$$\frac{\text{sale price}}{\text{decreased percent}} = \text{original price}$$

$$\frac{\$127.50}{0.85} = \frac{\$12750}{85} = \$150$$

Solution: The original price was $150.

Sample Problem 3

A man saved $225 on a new office desk that was on sale for 30% off. What was the original price of the desk?

Step 1: Let x indicate the original price. The amount of money the man saved ($225) is the difference between the original price and the sale price:

Original price – Sale price = $225

Step 2: The formula that helps to find the original price x is:

$x - p\% \cdot x$ = Sale price

We can rewrite this formula so that it's similar to the expression in Step 1 by getting the term "sale price" on the left with a minus sign, and getting the term $p\% \cdot x$ on the right side with a positive sign:

x – Sale price = $p\% \cdot x$

Original price – Sale price = $225

We can see now that the term $p\% \cdot x$ equals the amount of money the man saved.

A WORD OF ADVICE

The term $p\% \cdot x$ (where x is the original price) in the discount formula equals the amount of money saved after applying the discount.

Step 3: Since the term $p\% \cdot x$ equals the amount of money the man saved, we can express this mathematically as:

$p\% \cdot x = \$225$

Step 4: Change the percent of discount to decimal form:

30% = 0.30

Step 5: Substitute the percent of discount in decimal form into the equation in Step 3:

$0.30x = \$225$

Multiply both sides by 100 to eliminate decimals:

$30x = \$22{,}500$

To solve, divide both sides by 30:

$x = \$750$

Solution: The original price of the office desk was $750.

Retail stores buy their goods at wholesale prices and then add an amount called *markup* to cover overhead and make a profit. The markup is expressed as a percentage of the wholesale (original) price. The result is the retail price. This is somewhat backward compared to the discount problems. Since the sale price increases in comparison to the original price, the formula for finding the original price must be modified to reflect the fact that we need to add the difference in prices instead of subtract it, as is the case with a discount.

DEFINITION

Markup is expressed as a percentage of the wholesale price. It is added to the wholesale price to obtain the retail price.

If we know the sale price (often called retail price) and markup $p\%$, then the original price x (often called wholesale price) can be calculated using the following formula:

$x + p\% \cdot x = $ Sale price

Sample Problem 4

A shoe store uses a 40% markup on the cost of leather shoes. What is the wholesale cost of a pair of shoes that sells for $77?

Step 1: Let the wholesale (original) cost be x. We know that the retail price is $77. We also know that the markup is 40%.

Step 2: Change percent to the decimal form:

40% = 0.40

Step 3: Use the formula:

$x + p\% \cdot x$ = Sale price

Plug in the known values from Steps 1 and 2 into the formula:

$x + 0.40x$ = $77

$1.4x$ = $77

Divide both sides by 1.4:

x = $55

Solution: The wholesale (original) price was $55.

Finding Discount Price

When the problem asks us to find the discount (sale) price after discount, we can calculate the discount price using two methods again.

Method 1:

Step 1: Subtract the percent of discount from 100%:

100 percent – Percent of discount = Decreased percent

Step 2: Multiply the result by the original price to find the discount price:

(Original price) · (Decreased percent) = Discount (sale) price

Method 2:

Step 1: Multiply the original price by the percent of discount in decimal from to obtain the decrease in amount:

(Original price) · ($p\%$) = Decrease in amount (saved money)

Step 2: Subtract the decrease in amount from the original price to find the sale price:

Sale price = (Original price) – (Original price) · ($p\%$)

Obviously, using either method provides the same answer.

Sample Problem 5

The Howard University bookstore gives 15% discounts to students. What does a student actually pay for a book costing $24?

Let's use the first method to solve this problem.

Step 1: Let x indicate the sale price.

Step 2: Find the decreased percent by subtracting the percent of discount from 100%:

100% − 15% = 85%

Step 3: Let's change the decreased percent to decimal form:

85% = 0.85

Step 4: Find the sale price x by multiplying the original price by the decreased percent:

x = ($24)(0.85) = $20.40

Solution: The sale price of the book is $20.40.

DEAD ENDS

Always change the percent to decimal form; otherwise, you won't be able to solve the problem correctly.

Sample Problem 6

Mindy bought a color TV for 20% off the original price. What did she pay for the TV if the original price was $650?

Let's solve this problem using method 2.

Step 1: Let x indicate the discount price. Change the discount percent to the decimal form:

20% = 0.20

Step 2: Multiply the original price by the percent of discount in decimal form to obtain the decrease in the amount:

(Original price) · (p%) = Decrease in the amount (saved money)

$650 · 0.20 = $130

We found that Mindy saved $130.

Step 3: Subtract the savings from the original price to find the sale price:

(Original price) – (Saved amount) · ($p\%$) = Sale price

$650 – $130 = $520

Solution: The sale price for the TV was $520.

Finding the Percent of Discount

When the problem asks you to find the percent of discount, use the following formula:

$$\frac{\text{discount amount}}{\text{original price}} = \text{discount percent}$$

Sometimes the problem gives the discount amount (saved money); sometimes you need to find it by subtracting the sale price from the original price.

Sample Problem 7

At an office supply store, customers are given a discount if they buy more than $100 worth of goods. If a customer is given a discount of $19.32 on a total order of $552, what is the percent of discount?

Step 1: Let the discount percent be p. We know the discount amount, which is $19.32, and we know the original price, so we can use the formula right away:

$$p = \frac{\text{discount amount}}{\text{original price}} = \frac{\$19.32}{\$552}$$

$$\frac{\$19.32}{\$552} = 0.035$$

DEAD ENDS

The formula for the discount percent provides the percent in its decimal form. Don't forget to change the decimal form to percent; otherwise, the answer will be incorrect.

Step 2: We found the discount percent in its decimal form. Let's change it to a percent using the technique discussed earlier in Chapter 7:

0.035 = 3.5%

Solution: The discount percent was 3.5%.

Sample Problem 8

A $120 dress is on sale for $96. What is the discount percentage?

Step 1: Let p indicate the discount percent that we need to find. Next, let's find the discount amount (saved money) by subtracting the sale price from the original price:

$120 – $96 = $24

We found that the discount amount is $24.

Step 2: We know the original price and the discount amount, so we can use the formula to find the discount percent:

$$p = \frac{\text{discount amount}}{\text{original price}} = \frac{\$24}{\$120} = 0.2$$

Step 3: We found the discount percent in its decimal form. Let's change it to a percent:

0.2 = 20%

Solution: The discount percentage is 20%.

Double Discount Sales

At the end of the season, many retail stores apply an additional discount to already discounted items. For example, a 60% discount is followed by a 30% discount. Is this the same as putting an item at 90% discount, since 60% plus 30% is 90%?

Let's try to answer this question by solving two problems and comparing the solutions.

Sample Problem 9

A retail store put a cashmere pullover on 90% sale. If the original price was $200, what is the sale price?

Step 1: Let x indicate the sale price. Find the decreased percent by subtracting 90% from 100%:

100% – 90% = 10%

Step 2: Convert the decreased percent into its decimal form:

10% = 0.10

Step 3: Find the sale price by multiplying the original price by the decreased percent in the decimal form:

x = $200(0.10) = $20

Solution: The sale price is $20.

We found that the cost of the item with a straight 90% discount is $20.

Sample Problem 10

A retail store put a cashmere pullover on 60% sale. At the end of the season, the store gives an additional 30% discount. What is the sale price of the pullover at the end of the season if the original price was $200?

Step 1: Let x indicate the sale price after the 60% discount. First, find the decreased percent:

100% – 60% = 40%

Step 2: Change the decreased percent to its decimal form:

40% = 0.40

Step 3: Find the sale price by multiplying the original price by the decreased percent in the decimal form:

x = $200(0.40) = $80

We found that the sale price after the first 60% discount is $80. This amount becomes our new original price, which represents a new 100%.

DEAD ENDS

When a double discount is applied, adding the discount percents and then calculating the sale price is the wrong way to solve the problem.

Step 4: Let y indicate the new sale price. Find the decreased percent after the second discount by subtracting the additional discount percent (30%) from 100%:

100% − 30% = 70%

Step 5: Change the decreased percent into its decimal form:

70% = 0.70

Step 6: Find the sale price y by multiplying the new original price ($80) by the decreased percent in decimal form:

$y = \$80(0.70) = \$56$

Solution: The sale price is $56.

This price is quite different from the $20 price with a straight 90% discount.

Practice Problems

Problem 1: A suit was sold for $690 after a 25% discount. What was the regular price of the suit?

Problem 2: A coffee maker is on sale for $35. If the coffee maker is marked down 30%, what was the original price?

Problem 3: A bookstore has its regularly priced $45 book on art for 16% off the regular price. What is the sale price of the book?

Problem 4: How much will a living room set sell for if the regular price is $2,520 and the store gives a 15% discount?

Problem 5: All items at a tool store are marked up 25%. If the retail price of one tool set is $16.25, what was the wholesale cost to the owner?

Problem 6: An item that regularly sells for $850 is on sale for $637.50. What is the discount rate?

Problem 7: A retail store puts a $180 business suit on 60% discount. At the end of the season, the store gives an additional 15% discount. What is the sale price of the business suit at the end of the season?

The Least You Need to Know

- The general formula to find the original price x with the percent of discount $p\%$ and known sale price is: $x - p\% \cdot x = $ Sale price.

- The general formula to find the discount (sale) price when the original price and the discount percent are known is: Discount price = (Original price) − (Original price) $\cdot$ ($p\%$).

- Multiplying both sides of the equations by 100 helps eliminate decimals.

- To find the percent of discount, divide the discount amount (saved money) by the original price.

Geometry, Physics, and Liquid Solution Problems

In modern times, mathematics surrounds us in every aspect of our lives. Biology, medicine, economics, sports, the entertainment industry—all these fields of human activities use mathematics constantly. Often, mathematics applies rules and techniques from one of its branches to solve problems in another mathematical field. In this part, we first explore how algebra can help solve some geometry problems. Then we apply algebra rules to work through science-related problems that deal with concentrations of liquid solutions, temperature scales, and the lever principle.

When Algebra Helps Geometry

In This Chapter

- Finding measures of complementary and supplementary angles
- Mathematically expressing a missing angle in a triangle
- Calculating perimeters of polygons
- Computing the change in the area of a polygon when its dimensions are changed

Geometry word problems involve angle relationships, areas and perimeters of polygons, and properties of triangles and other polygons. If the illustration for the problem isn't provided, draw your own. Then fill in all the given information from the problem, such as measures of angles and lengths of sides, and represent the unknown quantities using variables. As the next step, use the geometry theorems and definitions to solve for unknown quantities, step by step, until you find the one the problem is asking for. You can refer to Appendix C for the list of geometry formulas we use in this chapter. In this chapter, we first work through problems to find unknown angle measures, then we calculate and compare perimeters of polygons. Finally, we learn how to find areas, including the area of a picture frame.

Cornering Angles

In the first two problems, we deal with *complementary* and *supplementary angles.* The strategy is to present one angle using a variable, then express the other one using given relationships, and, finally, apply geometry formulas and rules.

Sample Problem 1

If one of two supplementary angles is 5 more than 4 times the other, find the two angles.

Step 1: Let x represent the measure of the first smaller angle. Then the measure of its supplement can be expressed as $4x + 5$ (5 more than 4 times the other).

> **DEFINITION**
>
> **Complementary angles** are two angles whose measures have the sum 90°.
> **Supplementary angles** are two angles whose measures have the sum 180°.

Step 2: Using variables and values from Step 1, draw a sketch of the problem:

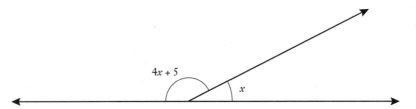

Step 3: Using the variables from Step 1, we can express the fact that the sum of two supplementary angles is 180° as:

$x + (4x + 5) = 180°$

Isolate the variable:

$x + 4x = 180 - 5$

Collect like terms on both sides:

$5x = 175$

Divide both sides by 5:

$x = 35°$

We found that the first angle is 35°.

A WORD OF ADVICE

In solving any geometry word problem, the sketch that has been drawn for you or a sketch you draw is the key to the answer.

Step 4: To find its supplement, substitute the found value for the first angle (35°) into the expression from Step 1:

$4x + 5 = 4(35) + 5 = 140 + 5 = 145°$

Alternatively, since the two angles are supplementary, we can find the supplement of the first angle by subtracting the measure of the first angle from 180°:

$180° - 35° = 145°$

Both methods provided the same answer.

We found that the measure of the supplement is 145°.

Solution: The two angles are 145° and 35°.

Sample Problem 2

One of two complementary angles is 3 less than twice the other angle. Find the two angles.

Step 1: Let x represent the measure of the second angle. Then the measure of the first angle can be expressed as $2x - 3$ (3 less than twice the other).

Step 2: Using variables from Step 1, draw a sketch of the problem:

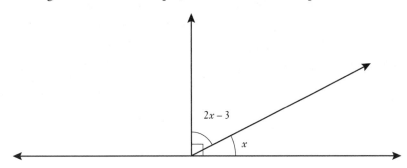

Step 3: Using the variables from Step 1, we can express the fact that the sum of two complementary angles is 90° as:

$x + (2x - 3) = 90°$

Isolate the variable:

$x + 2x = 90 + 3$

Collect like terms:

$3x = 93$

Divide both sides by 3:

$x = 31°$

We found that the second angle is 31°.

Step 4: To find the first angle, substitute the found value for the second angle (31°) into the expression from Step 1:

$2x - 3 = 2(31) - 3 = 62 - 3 = 59°$

We can also find the measure of the first angle by subtracting 31° from 90°, since the two angles are complementary:

$90° - 31° = 59°$

We found that the measure of the first angle is 59°.

Solution: The two angles are 59° and 31°.

For the next two problems, we use the fact that the measures of any triangle's three angles add up to 180°.

Sample Problem 3

In triangle PQR, the measure of $\angle P$ is 10 more than twice the measure of $\angle R$. The measure of $\angle Q$ is 4 less than 3 times the measure of $\angle R$. Find the measure of $\angle Q$.

Step 1: Let the measure of $\angle R$ be x, since the other angles are compared to this angle. Then the measure of $\angle P$ can be expressed as $2x + 10$, since it is 10 more than twice $\angle R$. The measure of $\angle Q$ can be expressed as $3x - 4$, since it is 4 less than 3 times the measure of $\angle R$.

Step 2: Using the variables and values we found in Step 1, draw a sketch of the triangle:

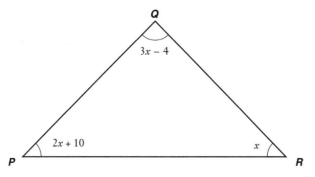

Step 3: Using the geometry fact that the sum of all angles in a triangle is equal to 180°, we can write the statement:

$\angle R + \angle P + \angle Q = 180°$

Substitute the values for each angle from Step 1:

$x + (2x + 10) + (3x - 4) = 180$

Open parentheses and isolate the variable:

$x + 2x + 10 + 3x - 4 = 180$

$x + 2x + 3x = -10 + 4 + 180$

Collect like terms on both sides:

$6x = 174$

Divide both sides by 6:

$x = 29$

We found that the measure of $\angle R$ is 29°.

Step 4: To find the measure of $\angle Q$, substitute the value for $\angle R$ into the expression from Step 1:

$\angle Q = 3x - 4 = 3(29) - 4 = 83°$

Solution: The measure of $\angle Q$ is 83°.

Sample Problem 4

In $\triangle PQR$, how many degrees are in $\angle QPR$?

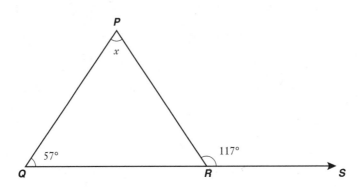

Step 1: Let the measure of $\angle QPR$ be x. To find x, we need to find the measure of $\angle PRQ$, since the measure of $\angle PQR$ is already known (57°).

A WORD OF ADVICE

Note that when an angle is named using three letters, the middle letter is the vertex of the angle.

Step 2: Using the geometry fact that the sum of all angles in any triangle is equal to 180°, we can write the statement:

$\angle QPR + \angle PRQ + \angle RQP = 180°$

Substituting the variables and known values, we can write:

$x + \angle PRQ + 57° = 180°$

Isolate the variable and collect like terms:

$x = 180° - \angle PRQ - 57°$

$x = 123° - \angle PRQ$

The last mathematical expression means that, to find x, we need to find the measure of $\angle PRQ$.

Step 3: To find the measure of $\angle PRQ$, we use the fact that two angles, $\angle PRQ$ and $\angle PRS$, are supplementary angles and that their sum is 180°:

$\angle PRQ + \angle PRS = 180°$

We know the measure of $\angle PRS$ (117°), so we can state:

$\angle PRQ + 117° = 180°$

We can find $\angle PRQ$ by subtracting 117° from 180°:

$\angle PRQ = 180° - 117° = 63°$

We found that the measure of $\angle PRQ$ is 63°.

Step 4: Now we can calculate x by substituting the value for $\angle PRQ$ (63°) into the last expression in Step 2:

$x = 123° - \angle PRQ$

$x = 123° - 63° = 60°$

Solution: The measure of $\angle QPR$ is 60°.

Measuring Perimeters

In the first two problems of this section, the relationships between the polygon's dimensions are given. We need to express these relations algebraically and then use geometry formulas to find the *perimeter.*

DEFINITION

The **perimeter** of a polygon is the sum of the lengths of the sides of the polygon.

Sample Problem 5

A triangle has a perimeter of 56. Two sides are equal, and the third is 5 more than the equal sides. Find the length of the third side.

Step 1: Let's first assign variables and let x represent the length of the equal sides. Then the length of the third side can be expressed as $(x + 5)$, since it is 5 more than the equal sides.

Step 2: Using variables from Step 1, make a sketch of the problem:

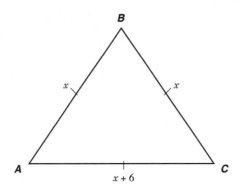

Step 3: We use the formula for a triangle's perimeter:

P = Sum of all 3 sides

Substitute the values from Step 1 and from the problem (for the perimeter):

$56 = x + x + (x + 5)$

Combine like terms on the right:

$56 = 3x + 5$

Isolate the variable:

$56 - 5 = 3x$

$3x = 51$

Divide both sides by 3:

$x = 17$

We found that the length of the equal sides is 17.

Step 4: To find the length of the third side, use the expression for the third side from Step 1 and substitute the value for the equal sides:

$x + 5 = 17 + 5 = 22$

Solution: The length of the third side is 22.

Sample Problem 6

The length of a rectangle is 4 feet less than twice the width. The perimeter of the rectangle is 34 feet. Find the dimensions of the rectangle.

Step 1: Let's first assign variables and allow x to represent the width of the rectangle. Then the length can be expressed as $(2x - 4)$, since it is 4 feet less than twice the width.

Step 2: Using variables from Step 1, make a sketch of the problem:

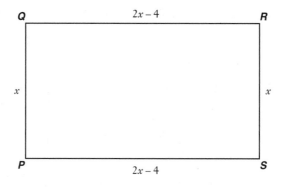

Step 3: We can use the formula for a rectangle's perimeter:

$P = 2L + 2W$

Substitute the values from Step 1 and from the problem (for the perimeter):

$34 = 2x + 2(2x - 4)$

Distribute on the right side:

$34 = 2x + 4x - 8$

Collect like terms on the right:

$34 = 6x - 8$

Isolate the variable:

$6x = 34 + 8$

$6x = 42$

Divide both sides by 6:

$x = 7$

We found that the width of the rectangle is 7 feet.

Step 4: To find the length, substitute the found value for the width into the expression from Step 1:

$2x - 4 = 2(7) - 4 = 14 - 4 = 10$

We found that the length is 10 feet.

Solution: The rectangle's dimensions are 10 feet and 7 feet.

The next problem deals with a perimeter as well. The difference is that a given polygon is transformed into another polygon, and we're dealing with the perimeters of two geometric shapes. But the routine is still the same: choose a variable, express other measures using algebra, and apply geometry formulas to find an unknown.

Sample Problem 7

If two opposite sides of a square increase by 6 and the other two opposite sides of the square decrease by 5, the perimeter of the resulting rectangle is 50. Find the perimeter of the square.

WORTHY TO KNOW

Squares and rectangles are special cases of polygons called quadrilaterals, which are four-sided polygons. Other special cases include parallelograms and trapezoids.

Step 1: Let x represent the side of the original square. Then the length of one pair of the rectangular sides will be $(x + 6)$, since we increased the corresponding square sides by 6. The length of the other pair of the rectangular sides can be expressed as $(x - 5)$, since we decreased the corresponding square sides by 5.

Step 2: Using variables from Step 1, make a sketch of the problem:

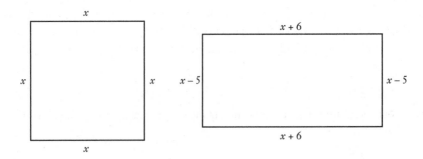

Step 3: We use the formula for a rectangle's perimeter and substitute variables and expressions from Step 1 and the known value for the perimeter:

$P = 2L + 2W$

$50 = 2(x + 6) + 2(x − 5)$

Distribute on the right side:

$50 = 2x + 12 + 2x − 10$

Collect like terms:

$50 = 4x + 2$

Isolate the variable:

$4x = 48$

Divide both sides by 4:

$x = 12$

We found that the side of the square is 12.

Step 4: We can use the formula for the square perimeter and substitute the found value for the side:

$P = 4L$

$P = 4(12) = 48$

Solution: The square perimeter is 48.

Covering Areas

In this section, we calculate the *areas* of polygons. Many people confuse perimeters with areas. A good idea here is to think about a rope that you can use, and mentally go around a polygon's perimeter to measure it. With areas, think about a piece of a carpet or a rug that you can use to cover a given area. Another helpful point to keep in mind is that you can go around the perimeter, but you cannot pave it with tiles.

DEFINITION

Area is the measure, in square units, of the interior region of a two-dimensional object.

Sample Problem 8

If two opposite sides of a square are each increased by 6 feet and the other two sides are each decreased by 2 feet, the area is increased by 60 square feet. Find the side of the square.

Step 1: Let's first deal with a square and let x represent its side. Then we need to find the area of the square, since the problem compares the areas of the rectangle and the square. According to the geometry formula for the area of a square, the square's area that we denote as A_1 can be expressed as:

$A_1 = x^2$

We found the area of the square.

Step 2: Now we can deal with a rectangle that has been formed by increasing one pair of the square sides and by decreasing the other pair. The measure of the longer side l can be expressed as $(x + 6)$, since it increased by 6 feet. The measure of the shorter side w can be expressed as $(x - 2)$, since it decreased by 2.

Step 3: Using variables from Steps 1 and 2, make a sketch of the problem:

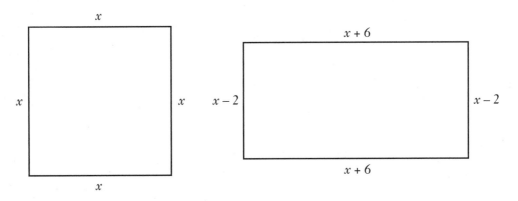

Step 4: The area of the rectangle that we denote as A_2 can be expressed, according to the formula, as:

$A_2 = w \cdot l = (x + 6)(x - 2)$

Multiply on the right side using the *FOIL method:*

$A_2 = x^2 + 6x - 2x - 12$

Collect like terms:

$A_2 = x^2 + 4x - 12$

We found the area of the rectangle.

A WORD OF ADVICE

When multiplying two binomials (two-term algebraic expressions), use the **FOIL method:**

$(x + 4)(x + 2) = x^2 + 4x + 2x + 8 = x^2 + 6x + 8$

Multiply the **f**irst terms: $x \cdot x$

Multiply the **o**uter terms: $2 \cdot x$

Multiply the **i**nner terms: $4 \cdot x$

Multiply the **l**ast terms: $4 \cdot 2$

Collect like terms.

Step 5: Now we use the problem's fact that the area of the rectangle is 60 square feet more than the area of the square, and express this fact mathematically as:

The area of the rectangle minus the area of the square equals 60.

$A_2 - A_1 = 60$

Substitute the two expressions for areas from Steps 1 and 4:

$x^2 + 4x - 12 - x^2 = 60$

Collect like terms on the left:

$4x - 12 = 60$

Note that the terms x^2 and $-x^2$ got cancelled.

Isolate the variable:

$4x = 60 + 12$

Collect like terms:

$4x = 72$

Divide both sides by 4:

$x = 18$

We found the side of the square.

Solution: The side of the square is 18 feet.

The next problem belongs to the most challenging type. This type deals with pictures in frames, sidewalks around buildings, and similar concepts. There are two key points to remember. The first is that twice the width of the frame, or the sidewalk, or additions around a swimming pool must be added to a polygon's dimensions, since the frame or the sidewalk goes all the way around a picture or a building. The second important thing is that the area of the frame (or anything else that goes around something) is the difference between the area of a picture with the frame and the area of the picture alone.

Sample Problem 9

The width of a painting without the frame is 7 inches shorter than its length. If the frame is 3 inches wide and its area is 174 square inches, find the dimensions of the painting alone.

Step 1: Since the width of the painting is compared to its length, let x represent the length of the painting without the frame. Then the width can be expressed as $x - 7$, since it is 7 inches shorter.

Step 2: The length of the painting with the frame can be calculated by adding 3 inches to the length of the painting without the frame 2 times (we have the frame on both sides):

$x + 3 + 3 = x + 6$

We found that the length of the painting with the frame is $x + 6$.

Step 3: The width of the painting with the frame can be calculated by adding 3 inches to the width ($x - 7$) of the painting without the frame 2 times (we have the frame on both sides as well):

$(x - 7) + 3 + 3 = x - 7 + 6 = x - 1$

We found that the width of the painting with the frame is $x - 1$.

WORTHY TO KNOW

Algebra can help solve many other geometry problems related to areas that would utilize the quadratic equation. You'll learn how to solve these problems while studying methods of solving quadratic equations.

Step 4: Using the variables and values we found in Steps 1, 2, and 3, make a sketch of the problem to help visualize the situation:

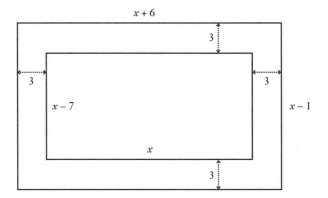

Step 5: Since the problem compares the areas of the painting with and without the frame, we need to express these areas. However, we have too many expressions from Steps 1, 2, and 3. It's helpful to organize them in a table to avoid confusion. The first column tells us whether we are dealing with the painting with or without the frame. The second column lists the corresponding widths. The third column displays the corresponding lengths. Finally, the fourth column records the corresponding areas that are found by multiplying the width by the length (based on the area formula for a rectangle $A = w \cdot l$):

	Width	**Length**	**Area**
Without frame	$x - 7$	x	$(x - 7)x$
With frame	$x - 1$	$x + 6$	$(x - 1)(x + 6)$

Step 6: Finally, we can use the problem's fact that the area of the frame is 174 square feet. Obviously, the area of the frame can be found by subtracting the area of the painting alone from the area of the painting with the frame (refer to our sketch). We can express this mathematically as:

The area of the painting with the frame minus the area of the painting without the frame equals the area of the frame.

$(x - 1)(x + 6) - (x - 7)x = 174$

Step 7: Multiply and distribute on the left side:

$x^2 + 6x - x - 6 - x^2 + 7x = 174$

Collect like terms:

$12x - 6 = 174$

Isolate the variable and collect free terms:

$12x = 174 + 6$

$12x = 180$

Divide both sides by 12:

$x = 15$

We found that the length of the painting alone is 15 inches.

Step 8: To find the width of the painting alone, substitute 15 inches into the expression for the width from Step 1:

$x - 7 = 15 - 7 = 8$

We found that the width of the painting alone is 8 inches.

Solution: The length of the painting alone is 15 inches, and the width of the painting alone is 8 inches.

Practice Problems

Problem 1: One of two complementary angles is 27 more than 8 times the other angle. Find the two angles.

Problem 2: If one of two supplementary angles is 5 less than 4 times the other, find the 2 angles.

Problem 3: The second angle of a triangle is 10° more than the first angle. The third angle is 5 more than 3 times the second angle. How many degrees are in each angle?

Problem 4: A triangle has a perimeter of 70. One side is 2 more than the other. The third side is 10 less than the other side. Find the length of all sides.

Problem 5: The length of a rectangle is 7 less than 3 times the width. The perimeter of the rectangle is 66 feet. Find the area of the rectangle.

Problem 6: If two opposite sides of a square are each increased by 12 inches, and the other two sides are each decreased by 8 inches, the area stays the same. Find the side of the square.

Problem 7: The length of a building is 15 feet less than twice its width. The sidewalk around the building is 8 feet wide. Find the dimensions of the building if the area of the sidewalk is 1,216 square feet.

The Least You Need to Know

- The measures of any triangle's three angles add up to 180°.
- To find the perimeter of any polygon, add up the lengths of all its sides.
- The area of a square is $A = a^2$, where a is the length of its side. The area of a rectangle is $A = w \cdot l$, where w is the width and l is the length.
- The area of a frame equals the difference between the area of a picture with the frame and the area of the picture alone.

Lever and Temperature Problems

In This Chapter

- Changing Fahrenheit into Celsius, and vice versa
- Converging Fahrenheit and Celsius temperature scales
- Balancing two or more weights
- Finding the distance from the fulcrum

Algebra helps us solve not only geometry-related word problems, but also many problems from the field of physics. In this chapter, we consider word problems pertaining to the lever principle and to the various temperature scales.

The Lever Principle

Lever problems are word problems that use the lever principle. Good examples of a lever are a seesaw, a teeterboard, a nutcracker, scissors, and a balance.

A lever can be set up with unequal weights placed at different distances from the balancing point, called the fulcrum. The distance from a weight to the fulcrum is called the arm of the weight.

To balance the lever, the sum of the products of the weights on one side of the fulcrum and their distances must equal the sum of the products of the weights on the other side of the fulcrum and their distances:

$$w_1 \cdot d_1 + w_2 \cdot d_2 = w_3 \cdot d_3 + w_4 \cdot d_4$$

$w_1, w_2, w_3,$ and w_4 are weights.

$d_1, d_2, d_3,$ and d_4 are corresponding distances from the fulcrum.

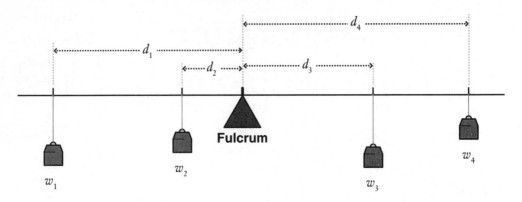

This general lever formula can be extended for more than two persons or weights on any side by adding more products for each person or weight.

Lever Problems

If there is only one weight on each side, the following formula is used:

$$w_1 \cdot d_1 = w_2 \cdot d_2$$

Sample Problem 1

Jim weighs 45 pounds and Sara weighs 30 pounds. They are both sitting on a seesaw. If Jim is seated 10 feet away from Sara, how far should each be from the fulcrum of the seesaw?

 A WORD OF ADVICE

Sketching a diagram of a problem situation is helpful in solving lever problems.

Step 1: Let d_1 stand for the distance of Jim from the fulcrum. Since the distance between the two children is 10 feet, then Sara's distance from the fulcrum is $d_2 = 10 - d_1$. Jim's weight is $w_1 = 45$ pounds, and Sara's weight is $w_2 = 30$ pounds.

Step 2: Using the variables and known values, sketch a diagram that helps to understand the problem:

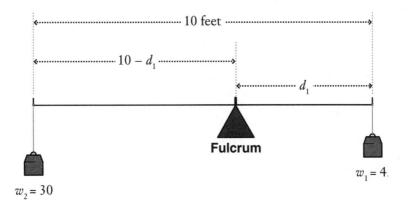

Step 3: Obtain the equation from the diagram:

$w_1 \cdot d_1 = w_2 \cdot d_2$

$45(d_1) = 30(10 - d_1)$

Distribute on the right side:

$45d_1 = 300 - 30d_1$

Isolate the variable and collect like terms:

$75d_1 = 300$

Divide both sides by 75:

$d_1 = 4$

We found that Jim should be 4 feet from the fulcrum.

Step 4: To find Sara's distance from the fulcrum, substitute 4 feet into the expression from Step 1:

$10 - d_1 = 10 - 4 = 6$

We found that Sara should be 6 feet from the fulcrum.

Solution: Jim should be 4 feet and Sara should be 6 feet from the fulcrum.

Sample Problem 2

Maria and Peter weigh 150 pounds together. They balance a teeterboard when Peter is 4 feet and Maria is 6 feet from the fulcrum. Find their weight.

Step 1: Since we need to find the children's weight and we know the combined weight, let x stand for Maria's weight. Then Peter's weight can be expressed as their combined weight minus Maria's weight: $150 - x$.

Step 2: Using the variables and known values, sketch a diagram that helps represent the problem:

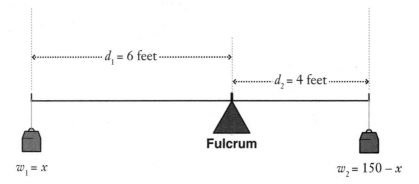

Step 3: Obtain the equation from the diagram:

$w_1 \cdot d_1 = w_2 \cdot d_2$

$x(6) = (150 - x)(4)$

Multiply on the left side and distribute on the right:

$6x = 600 - 4x$

Isolate the variable and collect like terms:

$6x + 4x = 600$

$10x = 600$

Divide both sides by 10:

$x = 60$ pounds

We found that Maria's weight is 60 pounds.

Step 4: To find Peter's weight, substitute 60 pounds into the expression for Peter's weight from Step 1:

$150 - x = 150 - 60 = 90$ pounds

We found that Peter's weight is 90 pounds.

Solution: Maria's weight is 60 pounds and Peter's weight is 90 pounds.

Sample Problem 3

Bob wants to move a 300-pound rock with a 6-foot, 6-inch crowbar. He puts the fulcrum 6 inches from the rock. How much force must he use to move the rock?

Step 1: Let x stand for the force Bob must use to move the rock. The distance from the fulcrum to Bob can be found by subtracting 6 inches (the distance from the fulcrum to the rock) from the length of the crowbar: 6 feet, 6 inches – 6 inches = 6 feet. Change 6 feet into inches using the fact that 1 foot is 12 inches:

6 feet = 6 · 12 = 72 inches

WORTHY TO KNOW

Archimedes, who extensively used lever principle in his inventions, once exclaimed, "Give me a place to stand and I will move the earth."

Step 2: Using the variables and known values, sketch a diagram that helps represent the problem:

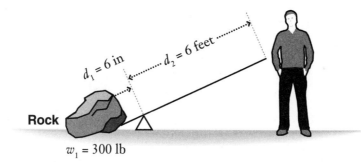

Step 3: Obtain the equation from the diagram:

$w_1 \cdot d_1 = w_2 \cdot d_2$

$w_1 \cdot d_1 = x \cdot d_2$

$300(6) = x(72)$

Multiply on both sides:

$1{,}800 = 72x$

Divide both sides by 72:

$x = 25$ pounds

Solution: Bob must use a force of 25 pounds.

When more than one person is sitting on one side from the fulcrum, use the general lever formula that we discussed at the beginning of the chapter.

Sample Problem 4

John, Lily, and Julia weigh 90, 70, and 60 pounds, respectively. John sits 4 feet, Lily sits 6 feet, and Julia sits 7 feet from the fulcrum on the same side. How far must their 200-pound father sit from the fulcrum to balance them all?

Step 1: Let x represent the distance of the father from the fulcrum.

Step 2: Using the variables and known values, sketch a diagram that helps represent the problem:

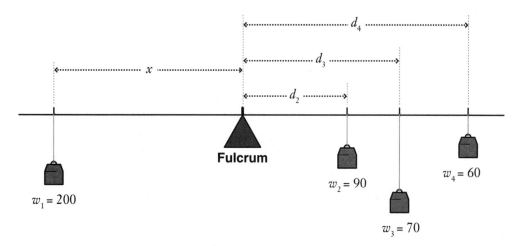

Step 3: Obtain the equation from the diagram:

$$w_1 \cdot d_1 = w_2 \cdot d_2 + w_3 \cdot d_3 + w_4 \cdot d_4$$

Plug in the known values and variables:

$$200x = (90)4 + (70)6 + (60)7$$

DEAD ENDS

When setting up an equation for lever problems with several people or weights on each side, be careful to use the correct corresponding distances; otherwise, the answer will be wrong.

Multiply on the right side:

$200x = 360 + 420 + 420$

Collect like terms:

$200x = 1,200$

Divide both sides by 200:

$x = 6$ feet

Solution: The father must sit 6 feet from the fulcrum.

Is It Hot or Cold?

Three scales are used to measure temperature: the Fahrenheit scale, the Celsius scale, and the Kelvin scale.

Fahrenheit is the temperature scale proposed in 1724 by, and named after, the physicist Daniel Gabriel Fahrenheit (1686–1736). Today, it is mostly used in the United States and a few other countries around the world for nonscientific purposes.

On the Fahrenheit scale, the freezing point of water is 32 degrees Fahrenheit (°F), and the boiling point is 212°F (at standard atmospheric pressure). There are 180 equal divisions between the freezing temperature of water and its boiling point. Each division is a Fahrenheit degree.

On the Celsius scale, the freezing point of water is 0 degrees Celsius (°C), and the boiling point of water is 100°C (at standard atmospheric pressure). There are 100 equal divisions between the freezing temperature of water and its boiling point. Each division is a Celsius degree.

Each degree Fahrenheit equals $\frac{5}{9}$ of a degree Celsius:

$$1°F = \frac{5}{9}°C$$

Each degree Celsius equals $\frac{9}{5}$ of a degree Fahrenheit:

$$1°C = \frac{9}{5}°F$$

Given the Fahrenheit reading, use the following formula to get the Celsius reading:

$$°C = \frac{5}{9}(F - 32)$$

Given the Celsius reading, use the following formula to get the Fahrenheit reading:

$$°F = \frac{9}{5}C + 32$$

> **WORTHY TO KNOW**
>
> Zero degrees on the Kelvin scale is called the absolute zero of temperature. Absolute zero is the coldest temperature theoretically possible. It cannot be reached by any artificial or natural means.

The Kelvin scale has the same divisions as the Celsius scale, except that the freezing point of water on the Kelvin scale is 273. Each division is denoted as 1°K.

To find the Kelvin reading, use the following formula:

°K = °C + 273

Temperature Problems

Temperature problems usually involve conversions from one temperature scale to another.

Sample Problem 5

When I lived in Russia, one of the coldest outside temperatures I experienced was –45°C. The coldest temperature I experienced in the United States (in Maryland) was +3°F. What is the difference in the coldest temperatures on the Fahrenheit scale?

Step 1: Change the Celsius temperature to Fahrenheit using the formula and substituting the value of –45°C:

$$°F = \frac{9}{5}C + 32$$

$$°F = \frac{9}{5}(-45) + 32$$

$$°F = \frac{9(-\overset{9}{\cancel{45}})}{\underset{1}{\cancel{5}}} + 32 = -81 + 32 = -49°F$$

We found that the coldest temperature I experienced in Russia was –49°F.

Step 2: To find the difference between the coldest temperatures in two countries, subtract the coldest temperature in Russia from the coldest temperature in the United States:

+3°F – (–49°F) = +3°F + 49°F = 52°F

Solution: The difference between the two coldest temperatures is 52°F.

Sample Problem 6

Find the temperature where both temperature scales converge.

Step 1: Since we want to equalize both temperature scales, we can write this mathematically as:

°C = °F

A WORD OF ADVICE

The two temperature scales converge only at one temperature: –40°C = –40°F.

Step 2: Use the conversion formula:

$$F = \frac{9}{5}C + 32$$

Plug in the right side of the preceding equation into the right side of the equation in Step 1:

$$C = \frac{9}{5}C + 32$$

Isolate the variable:

$$C - \frac{9}{5}C = 32$$

Multiply both sides by 5:

5C – 9C = 5(32)

Collect like terms and multiply on the right side:

–4C = 160

Divide both sides by –4:

C = –40

We found that a temperature of $-40°C = -40°F$ is the one at which both temperature scales converge.

Solution: $-40°C = -40°F$.

Sample Problem 7

Normal room temperature is 72°F. What is it on the Celsius scale?

Step 1: Change the Fahrenheit temperature to Celsius using the formula and substituting the value of 72°F:

$$C = \frac{5}{9}\left(F - 32\right)$$

$$C = \frac{5}{9}\left(72 - 32\right) = \frac{5}{9} \cdot 40 = \frac{200}{9} = 22.22$$

We rounded the answer to the hundredths place.

Solution: Normal room temperature is 22.22°C.

Sample Problem 8

If the temperature drops 36°F, what is the corresponding drop on the Celsius scale?

Step 1: We know that each degree Fahrenheit equals $\frac{5}{9}$ of a degree Celsius, so we can state:

$$1°F = \frac{5}{9}°C$$

Step 2: We know how much 1 Fahrenheit degree is. To find how much 36°F is, we need to multiply the right side of the expression in Step 1 by 36:

$$(36) \cdot \frac{5}{9} = \frac{\overset{4}{\cancel{36}} \cdot 5}{\underset{1}{\cancel{9}}} = 20°C$$

Solution: A 36°F drop of temperature on the Fahrenheit scale means a 20°C temperature drop on the Celsius scale.

Sample Problem 9

If the temperature raises 45°C, what is the corresponding rise on the Fahrenheit scale?

Step 1: We know that each degree Celsius equals $\dfrac{9}{5}$ of a degree Fahrenheit, so we can state:

$$1°C = \frac{9}{5}°F$$

Step 2: Since we know how much 1 Celsius degree is, to find how much 45°C is, we need to multiply the right side of the expression in Step 1 by 45:

$$(45)\frac{9}{5} = \frac{\overset{9}{\cancel{45}} \cdot 9}{\underset{1}{\cancel{5}}} = 81°F$$

Solution: A 45°C rise in temperature on the Celsius scale means an 81°F rise on the Fahrenheit scale.

Sample Problem 10

Find the corresponding Kelvin temperature reading of a 95°F reading.

Step 1: To find the Kelvin reading, we first need to find the Celsius reading. Change the Fahrenheit temperature to Celsius using the formula and substituting the value of 95°F:

$$C = \frac{5}{9}\left(F - 32\right)$$

$$C = \frac{5}{9}\left(95 - 32\right) = \frac{5}{9} \cdot 63 = \frac{315}{9} = 35$$

We found that 95°F is 35°C.

A WORD OF ADVICE

To change a Fahrenheit temperature into a Kelvin temperature, first convert the Fahrenheit reading into a Celsius reading, and then change it to Kelvin.

Step 2: To find the corresponding Kelvin temperature, use the formula and substitute the value of 35°C:

°K = °C + 273

K = 35 + 273 = 308°K

Solution: The corresponding Kelvin reading is 308°K.

Practice Problems

Problem 1: A 400-pound weight rests on one end of a lever, 24 feet from the fulcrum. What weight, resting on the opposite end and 6 feet from the fulcrum, would make the lever balance?

Problem 2: A gardener was preparing a new flower bed. Suddenly he hit a large rock and used a crowbar to move the rock. The crowbar was 8 feet long, and the rock was later measured to weigh 160 pounds. The gardener used another rock as the fulcrum and used a force equal to 32 pounds to raise the rock to a point of balance. How far was the fulcrum from the 160-pound rock?

Problem 3: Four boys decided to use the same teeter-totter. Two of them weighing 85 pounds and 55 pounds, respectively, sat on opposite ends of the 14-foot board, which had the balance point at the center. The third boy, weighing 60 pounds, sat on the same side as the one weighing 85 pounds and was 5 feet from the fulcrum. Where must the fourth boy sit on the same side as the one weighing 55 pounds so that the teeter-totter is balanced if the fourth boy weighs 102 pounds?

Problem 4: The room temperature is 77°F. What is it on the Celsius scale?

Problem 5: The oven temperature is 85°C. What is it on the Fahrenheit scale?

Problem 6: Find the corresponding Kelvin temperature reading of 95°F reading.

Problem 7: If the temperature drops 63°F, what is the corresponding drop on the Celsius scale?

The Least You Need to Know

- The main formula for lever problems is: $w_1 \cdot d_1 + w_2 \cdot d_2 = w_3 \cdot d_3 + w_4 \cdot d_4$.
- To convert Celsius degrees into Fahrenheit degrees, use the formula
  $$°F = \frac{9}{5}C + 32.$$
- To convert Fahrenheit degrees into Celsius degrees, use the formula
  $$°C = \frac{5}{9}(F - 32).$$
- Each degree Fahrenheit equals $\frac{5}{9}$ of a degree Celsius. Each degree Celsius equals $\frac{9}{5}$ of a degree Fahrenheit.

Liquid Solution Problems

In This Chapter

* Finding how much of one solution to add to another
* Removing water to make the solution stronger
* Replacing solutions to change concentration
* Making alloys of different concentration

Many human activities, including cooking, biology, medicine, and chemistry, involve mixing different substances and producing *liquid mixtures* of different strength. These are also called liquid solutions.

Solving liquid solution or liquid mixture word problems doesn't involve one main formula, but two important points factor into how you approach these problems. The first suggestion is to think in terms of a pure substance: a phrase such as "a 40% alcohol solution" means that 40% of the amount of solution is pure alcohol and the rest (60%) is water. The percentage of the solution that is alcohol plus the percentage of the solution that is water is equal to 100%, all of the solution. Generally, for an x% solution of two substances that are mixed, x% of the solution is one substance and $(100 - x)$% is the other substance.

The second suggestion is to think about the part–whole relationship that we discussed in Chapter 7. Suppose you need to find how much pure orange juice is in 2 gallons of a 30% orange juice solution. We need to find what is 30% of 2. As we've already learned, the answer is to multiply the percent in decimal form by the whole amount of the solution (2 gallons), so there is (0.30)(2), or 0.6, gallons of pure orange juice in 2 gallons of a 30% orange juice solution.

> **DEFINITION**
>
> **Liquid mixtures** have specific strength and are made from two or more solutions with a different concentration.

Let's emphasize these two suggestions one more time with an additional example. Suppose you mixed 5 gallons of a 50% antifreeze solution with 4 gallons of a 70% antifreeze solution. Find how much pure antifreeze is in the mixture. First change percents into decimal form: 50% = 0.50, 70% = 0.70. The number of gallons of pure antifreeze in the first solution is 0.50(5) = 2.5 gallons, and the number of gallons of pure antifreeze in the second solution is 0.70(4) = 2.8 gallons. The fact that the two solutions are mixed doesn't change the total amount of pure antifreeze, regardless of how it is distributed in the two solutions.

To solve the problem, we need to translate the following word guideline into mathematics language:

The amount of pure antifreeze in the original plus the amount of pure antifreeze to be added equals the amount of pure antifreeze in the final solution.

0.50(5) + 0.70(4) = Pure antifreeze

2.5 + 2.8 = 5.3

We found that there are 5.3 gallons of pure antifreeze in the mixture.

These key ideas will help you solve the liquid solution problems in this chapter. The last thing I want to point out is that the solution technique for liquid mixture problems is very similar to the technique we used for dry mixture problems.

Adding to the Solution

Suppose that we add together two sugar solutions. The only two substances involved are sugar and water. When we pour together two different solutions, we produce a new solution that is again made only of sugar and water. All the water in the new solution came from the water in the two solutions we mixed, and all the sugar in the new solution came from the same two solutions we mixed. Keep thinking along these lines while solving problems.

Sample Problem 1

How many gallons of a 70% acid solution must be added to 50 gallons of a 40% acid solution to produce a 50% acid solution?

Step 1: Since we need to find how many gallons of the first solution to mix, let x represent this amount. Then the amount of the resulting solution is $x + 50$, since we must add 50 gallons of the second solution. Also, convert acid percent in all solutions into decimal form:

70% = 0.70

40% = 0.40

50% = 0.50

Step 2: Using the variables and values from Step 1, let's set up a table with four columns and four rows. The first column shows the type of solution. The amount of each type of solution found in Step 1 and given by the problem (50 gallons) is listed in the second column. The third column displays the acid percent in decimal form. Finally, the fourth column displays the total amount of acid in gallons. The total amount is found by multiplying the acid percent by the amount of each solution type.

	Amount	Percent Acid	Total Gallons Acid
70% sol'n	x	0.70	$0.70x$
40% sol'n	50	0.40	$(0.40)(50)$
Mixture	$x + 50$	0.50	$(0.50)(x + 50)$

Step 3: The Total Gallons Acid column will be used to write the equation:

The acid in the 70% solution plus the acid in the 40% solution equals the total acid.

$0.70x + (0.40)(50) = (0.50)(x + 50)$

To solve the equation, multiply and distribute on both sides:

$0.70x + 20 = 0.50x + 25$

A WORD OF ADVICE

Creating a table for liquid mixtures problems helps to organize the understanding of the problem.

Multiply both sides by 100 to eliminate decimals:

$70x + 2,000 = 50x + 2,500$

Isolate the variable:

$70x - 50x = 2,500 - 2,000$

Collect like terms:

$20x = 500$

Divide both sides by 20:

$x = 25$ gallons

Solution: We must add 25 gallons of a 70% acid solution.

Sample Problem 2

A nurse mixed 96 ounces of a 4% iodine solution with 80 ounces of a 15% iodine solution. What is the percentage of iodine in the mixture?

Step 1: We know the amount in ounces of all solutions. We also know the strength of the two solutions that are being mixed. What we don't know is the percentage of iodine in the final solution. Let x represent this percentage.

Step 2: We also need to change percents into decimal form:

$4\% = 0.04$

$15\% = 0.15$

The last thing we need to do is add the amount of the two solutions that are mixed to find the amount in ounces of the final solution:

$96 + 80 = 176$ ounces

We found that the amount of the final mixture is 176 ounces.

Step 3: Using the variables and values we found in Steps 1 and 2, we can create a table to organize the problem. The first column shows the type of solution. The amount of each type of solution found in Steps 1 and 2 is listed in the second column. The third column displays the iodine percent in decimal form. Finally, the fourth column displays the total amount of iodine in ounces. The total amount is found by multiplying the iodine percent by the amount of each solution type.

	Amount	Percent Iodine	Total Ounces Iodine
4% sol'n	96	0.04	96(0.04)
15% sol'n	80	0.15	80(0.15)
Mixture	176	x	$176x$

Step 4: We use the fourth column to write the equation:

The iodine in the 4% solution plus the iodine in the 15% solution equals the total iodine.

$96(0.04) + 80(0.15) = 176x$

Multiply on the left side:

$3.84 + 12 = 176x$

Collect like terms:

$15.84 = 176x$

DEAD ENDS

Note that when we mix two solutions with different percentages, the percentage of the resulting solution must be in between these two percentage values. For example, if you mix a 14% solution with a 35% solution, the resulting solution can't have a percentage higher than 35% or lower than 14%; it must be between 14% and 35%.

Multiply both sides by 100 to eliminate decimals:

$1,584 = 17,600x$

Divide both sides by 17,600:

$x = 0.09$

We found the percentage of the iodine in decimal form. Change it to a percent:

$0.09 = 9\%$

Solution: The percentage of iodine in the final solution is 9%.

Sample Problem 3

A chemistry intern needs 20 liters of a 15% sodium solution. He decided to mix a 10% solution with a 30% solution. How many liters of a 10% solution and a 30% solution does he need?

Step 1: We know that the amount of the final solution is 20 liters. But we don't know the amount of the original solutions, so let x stand for the number of liters of a 10% solution, and let y stand for the number of liters of a 30% solution. Then the number of liters of the final solution can be expressed as $x + y$. At the same time, we know that the amount of the final solution is 20 liters, so we can state that $x + y = 20$.

Step 2: Let's also convert percents into decimal form:

10% = 0.10

30% = 0.30

15% = 0.15

Step 3: Using the variables and values we found in Steps 1 and 2, we can create a table to organize the problem. The first column shows the type of solution. The amount of each type of solution found in Step 1 is listed in the second column. The third column displays the sodium percent in decimal form. Finally, the fourth column displays the total amount of sodium in liters. The total amount is found by multiplying the sodium percent by the amount of each solution type.

	Amount	**Percent Sodium**	**Total Liters Sodium**
10% sol'n	x	0.10	(0.10)x
30% sol'n	y	0.30	(0.30)y
Mixture	$x + y = 20$	0.15	(0.15)20

Step 4: Since we have two variables, we use a system of two equations to solve the problem. The first equation states that the number of liters of the final solution is 20:

$x + y = 20$

We use the fourth column to write the second equation:

The sodium in the 10% solution plus the sodium in the 15% solution equals the total sodium.

$(0.10)x + (0.30)y = (0.15)20$

Then the system of two equations is:

$x + y = 20$

$(0.10)x + (0.30)y = (0.15)20$

Step 5: To solve the system, we revise the first equation and substitute for x into the second equation:

$x = 20 - y$

$(0.10)(20 - y) + (0.30)y = (0.15)20$

> **WORTHY TO KNOW**
>
> Many mixture problems can be solved using either one equation or a system of two equations. The solving method depends on the way we choose variables. If in problem 3 we choose variables in a way that x stands for the amount of a 10% solution, then the amount of a 30% solution can be expressed as 20 – x (since there are 20 liters of the final solution). In this case, the problem can be solved using just one equation.

Distribute and multiply:

$2 - 0.10y + 0.30y = 3$

Multiply by 100 to eliminate decimals:

$200 - 10y + 30y = 300$

Collect like terms and isolate the variable:

$20y = 100$

Divide both sides by 20:

$y = 5$ liters

Since y stands for the amount of a 30% solution, we found that the intern needs 5 liters of a 30% solution.

Step 6: To find the amount of a 10% solution, substitute y into the revised equation in Step 5:

$x = 20 - y = 20 - 5 = 15$ liters

We found that the intern needs 15 liters of a 10% solution.

Solution: The intern needs 15 liters of a 10% solution and 5 liters of a 30% solution.

Sometimes the solution that is added is pure water. It is obvious that the concentration of the solution to which water is added will become weaker. Let's do one problem like that and see how it works.

Sample Problem 4

How many ounces of pure water must be added to 150 ounces of a 15% salt solution to make a salt solution that is 10% salt?

Step 1: Since we don't know the amount of ounces of pure water to add, let x stand for this amount. Water will be added to 150 ounces of the original solution, so we can state that the amount of the final solutions is $x + 150$.

Step 2: Let's also change percents into decimal form:

15% = 0.15

10% = 0.10

Step 3: Using the variables and values we found in Steps 1 and 2, create a table to organize the problem. The first column shows the type of solution. The second column lists the amount of each type of solution found in Step 1. The third column displays the salt percent in decimal form. Finally, the fourth column displays the total amount of salt in ounces. The total amount is found by multiplying the salt percent by the amount of each solution type.

Note one important point: since pure water contains no salt, the percent of salt is 0, and the corresponding total amount of salt is also 0.

	Amount	Percent Salt	Total Ounces Salt
Water	x	0	0
15% sol'n	150	0.15	(0.15)(150)
10% mix	$x + 150$	0.10	(0.10)($x + 150$)

DEAD ENDS

In liquid mixture problems, the equation is usually about the amount of some substance such as salt, iodine, or sodium that is dissolved in another substance (usually in water). The absence of one term in the equation, when the amount of the first substance is 0 (like 0 salt in pure water), does not mean that we're not taking this into account in our equation. The amount of pure water is still incorporated into the equation as x.

Step 4: Note that all the salt in the resulting solution comes from the amount of salt in the original solution, since, as we discussed before, pure water has 0% salt. We can write the problem's equation using the last column:

The salt in a 15% solution equals the salt in the final solution.

$(0.15)(150) = (0.10)(x + 150)$

Multiply and distribute both sides:

$22.5 = 0.10x + 15$

Multiply by 100 to eliminate decimals:

$2{,}250 = 10x + 1{,}500$

Isolate the variable:

$750 = 10x$

Divide both sides by 10:

$x = 75$ ounces

Solution: Seventy-five ounces of pure water must be added to make the solution 10% salt.

Sample Problem 5

A jewelry maker mixed 1,000 ounces of a silver alloy with 400 ounces of another silver alloy, which is 35% less silver than the first. If the final alloy is 85% silver, what is the percentage of silver in each alloy?

Step 1: We know the weight of the first alloy (1,000 ounces) and the weight of the second alloy (400 ounces), thus the weight of the final alloy will be 1,000 + 400 = 1,400 ounces. What we don't know is the percentage of silver in each alloy. Let x stand for the percent of silver in the first alloy; then the percent of silver in the second alloy is $x - 35\%$, since the second alloy is 35% less silver.

Step 2: We also need to change percents into decimal form:

$35\% = 0.35$

$85\% = 0.85$

Then the percentage of the second alloy in decimal form is:

$x - 35\% = x - 0.35$

Step 3: Using the variables and values we found in Steps 1 and 2, create a table to organize the problem. The first column shows the type of alloy. The second column lists the amount of each type of alloy found in Step 1. The third column displays the silver percent in decimal form. Finally, the fourth column displays the total amount of silver in ounces. The total amount is found by multiplying the silver percent by the amount of each alloy type.

	Amount	Percent Silver	Total Ounces Silver
Alloy 1	1,000	x	$1,000x$
Alloy 2	400	$x - 0.35$	$400(x - 0.35)$
Final alloy	1,400	0.85	$(1,400)(0.85)$

Step 4: We can write the equation using the last column:

The silver in Alloy 1 plus the silver in Alloy 2 equals the silver in the final alloy.

$1,000x + 400(x - 0.35) = (1,400)(0.85)$

Distribute the left side and multiply on the right:

$1,000x + 400x - 140 = 1,190$

Collect like terms and isolate the variable:

$1,400x = 1,330$

Divide by 1,400:

$x = 0.95$

We found the percent of silver in the first alloy in decimal form. Change it to a percent:

$0.95 = 95\%$

We found that the silver percent in the first alloy is 95%.

Step 5: To find the percent of silver in the second alloy, substitute 95% into the expression from Step 1:

$x - 35\% = 95\% - 35\% = 60\%$

We found that the percent of silver in the second alloy is 60%.

Solution: The percent of silver in the first alloy is 95% and, in the second, 60%.

Removing from the Solution

Sometimes, to make the solution stronger, some amount of water must be evaporated. The technique for solving this type of problem is generally the same with some little nuances.

Sample Problem 6

A chemist has 40 ounces of a 20% sodium solution. How much water should be evaporated to make it a 30% solution?

Step 1: Let x represent the amount of water evaporated. Since 20% of the original solution was sodium, the original concentration of water was:

$100\% - 20\% = 80\%$

The resulting concentration of water is this, since 30% was sodium:

$100\% - 30\% = 70\%$

Step 2: Let's change the percent of water in the original and final solutions into decimals:

$80\% = 0.80$

$70\% = 0.70$

Step 3: Using the variables and values we found in Steps 1 and 2, create a table to organize the problem. But in this problem, the table will be about the solvent medium (water), not about the diluted material (sodium), as in the previous problems. The first column shows the type of the solution. The second column lists the amount of each type of solution found in Step 1. The third column displays the water percent in decimal form from Step 2. Finally, the fourth column displays the total amount of water in ounces. The total amount is found by multiplying the water percent by the amount of each solution type.

Note that the water evaporated is 100% water, which is just 1 in decimal form.

	Amount	Percent Water	Total Ounces Water
Original	20	0.80	(0.80)20
Removed	x	1	$1x$
Final sol'n	$20 - x$	0.70	(0.70)(20 − x)

Step 4: To set up the problem's equation, we need to translate the following statement into algebra language:

The total water in the original solution minus the water removed from the final solution equals the total water.

$20(0.80) - x = (0.70)(20 - x)$

Step 5: To solve the equation, multiply on the left side and distribute on the right side:

$16 - x = 14 - 0.70x$

Multiply both sides by 100 to eliminate decimals:

$1{,}600 - 100x = 1{,}400 - 70x$

Isolate the variable:

$-100x + 70x = 1{,}400 - 1{,}600$

Collect like terms:

$-30x = -200$

Divide both sides by −30 and round the answer to the hundredths place:

$x \approx 6.67$ ounces

Solution: The chemist should evaporate approximately 6.67 ounces of water.

DEAD ENDS

When, instead of evaporation, pure water is added to a solution, we need to add, not subtract, the amount of pure water to the original solution amount to write the problem's equation.

Sample Problem 7

If 8 liters of water are removed from 40 liters of a 20% salt solution, what will be the percentage of salt in the resulting solution?

Step 1: Let x represent the percentage of salt in the resulting solution. Since 20% of the original solution was salt, the original concentration of water was:

100% – 20% = 80%

Also, if x% of the resulting solution is salt, then the concentration of water in the resulting solution will be 100% – x, or 1 – x in decimal form.

Step 2: Let's change the percent of water in the original solution to a decimal:

80% = 0.80

Eight liters of water that are removed are 100% water, which is just 1 in the decimal form.

We also can calculate that the amount of the final solution is 40 – 8 = 32 liters, since the amount of the original solution is 40 liters and the amount of the removed water is 8 liters.

Step 3: Using the variables and values we found in Steps 1 and 2, create a table to organize the problem. Again, in this problem, the table will be about the solvent medium (water), not about the diluted material (salt), as in the previous problems. The first column shows the type of solution. The second column lists the amount of each type of solution found in Step 1 and given by the problem. The third column displays the water percent in decimal form from Step 2. Finally, the fourth column displays the total amount of water in liters. The total amount is found by multiplying the water percent by the amount of each solution type.

	Amount	Percent Water	Total Liters Water
Original	40	0.80	40(0.80)
Removed	8	1	8
Final sol'n	32	1 – x	32(1 – x)

Step 4: To set up the problem's equation, we need to translate the following statement into algebra language:

The total water in the original solution minus the water removed from the final solution equals the total water.

$$40(0.80) - 8 = 32(1 - x)$$

Step 5: To solve the equation, multiply on the left side and distribute on the right:

$$32 - 8 = 32 - 32x$$

Isolate the variable:

$$32 - 32 - 8 = -32x$$

Collect like terms:

$$-8 = -32x$$

Divide both sides by -32:

$$x = 0.25$$

We found that the percentage of salt of the resulting solution in decimal form is 0.25. Change this to a percent:

$$0.25 = 25\%$$

Solution: The percentage of salt in the resulting solution is 25%.

Replacing the Solution

Sometimes, to change the strength of the solution, some amount of the original solution must be removed and then replaced by a new solution. The key point to understand is that the amount of the solution removed is equal to the amount of the solution that replaces it.

Sample Problem 8

A tank contains 20 gallons of antifreeze solution. When it is full, it contains 15% antifreeze. How many gallons must be replaced by an 80% antifreeze solution to get 20 gallons of a 70% solution?

Step 1: Let x represent the amount of antifreeze solution replaced. This means that we first removed x gallons of a 15% solution and then added x gallons of an 80% solution to make the resulting solution 70% antifreeze. The amount of the resulting solution is the same as the original solution's amount, which is 20 gallons.

A WORD OF ADVICE

When replacing solutions, the amount of the removed solution is usually equal to the amount of the added solution.

Step 2: Let's now change percents to decimals:

Original   15% = 0.15

Removed 15% = 0.15

Added   80% = 0.80

Resulting 70% = 0.70

Note that the removed solution obviously has the same strength as the original one.

Step 3: Using the variables and values we found in Steps 1 and 2, create a table to organize the problem. This time, the table needs an additional row for the added solution. The first column shows the type of solution. The second column lists the amount of each type of solution found in Step 1 and given by the problem. The third column displays the antifreeze percent in decimal form from Step 2. Finally, the fourth column displays the total amount of antifreeze. The total amount is found by multiplying the antifreeze percent by the amount of each solution type.

	Amount	Antifreeze %	Total Gallons Antifreeze
Original	20	0.15	20(0.15)
Removed	x	0.15	0.15x
Added	x	0.80	0.80x
Final sol'n	20	0.70	20(0.70)

Step 4: To set up the problem's equation, we need to translate into algebra language the following statement:

Antifreeze in the original solution minus antifreeze in the removed solution plus antifreeze in the added solution equals antifreeze in the resulting solution.

$20(0.15) - 0.15x + 0.80x = 20(0.70)$

Step 5: To solve the equation, multiply on both sides first:

$3 - 0.15x + 0.80x = 14$

Multiply each term by 100 to eliminate decimals:

$300 - 15x + 80x = 1,400$

Collect like terms and isolate the variable:

$65x = 1,400 - 300$

$65x = 1,100$

Divide both sides by 65:

$x \approx 16.92$ gallons

We rounded the answer to the hundredths place.

Solution: A total of 16.92 gallons of a 15% antifreeze solution need to be replaced.

Practice Problems

Problem 1: How many ounces of a 35% acid solution and a 10% acid solution need to be mixed to make 30 ounces of a 20% solution?

Problem 2: A scientist mixed 4 liters of a 30% sodium solution and 6 liters of a 50% sodium solution. What is the sodium percentage in the final mixture?

Problem 3: A chemist mixed a 20% alcohol solution with another 60% alcohol solution to make a 34% alcohol solution. If there are 36 ounces less of the 60% alcohol solution than of the 20% alcohol solution, how many ounces are there in the total mixture?

Problem 4: How many liters of pure water must be added to 300 liters of a 30% acid solution to make an acid solution that is 20% acid?

Problem 5: A metallurgist needs to make 24.8 ounces of an alloy that is 50% gold. He is going to melt and combine one alloy that is 60% gold with another alloy that is 40% gold. How much of each alloy should he use?

Problem 6: How much water must be removed from 90 ounces of a 15% sugar solution to make a solution that is 20% sugar?

Problem 7: A tank contains 48 gallons of a 20% antifreeze solution. How much solution needs to be removed and replaced with pure antifreeze to get a 40% solution?

The Least You Need to Know

- When creating a table to organize the problem, be very clear whether you are doing the table for a diluted material or a solvent medium (usually water).

- In the table that we create to organize the problem, the last column displays the total amount of a diluted material. Usually the last column of the table helps to write the equation.

- Removing water from a solution makes the solution stronger; adding water makes it weaker.

- When replacing a solution, first subtract the removed amount from the amount of the original solution, and then add the amount of the replacing solution in order to write the equation.

Average and Motion Problems

The last part of the book puts your algebra skills to the ultimate test. First, we work through average problems and learn to calculate the minimum score you can get on a test and still earn a good overall average in the class. You also learn here how an excellent grade for a "heavy-weighted" four-credit class helps in securing a great GPA.

Later, in two action-packed motion chapters, we discuss various scenarios of traveling: traveling with an average speed, catching up, moving toward each other, traveling with or against the wind, and much more. Writing algebra equations for these problems is often a challenge, but you can do it!

In the final chapter of the book, we summarize the strategies and approaches we've learned and apply them to problems that defy definition—that is, problems we can't clearly classify as any type of problem we've discussed thus far.

Chapter 17

Is Arithmetic Mean?

In This Chapter

- Finding the removed or added term
- Calculating the least score to sustain a certain average
- Obtaining the sum of weighted terms
- Computing the GPA

If you were to pick a number that best describes all the heights of players on your soccer team, what number would you pick? You would probably put all the heights in order from least to greatest and choose a number somewhere in the middle of the data. This number is the average. There are three types of average: mean, median, and mode. In this chapter, we deal with the arithmetic mean. We learn how to use the formula for the mean and find a missing member of the data set—for example, an unknown test score. Later, we explore the weighted average and its use in computing grades and GPA.

What Is Arithmetic Mean?

The *arithmetic mean* or average uses the following formula:

$$\text{Average} = \frac{\text{Sum of Terms}}{\text{Number of Terms}}$$

Frequently, all that an "average" problem wants from you is to go back and forth a few times between totals (sum of all terms) and averages. But most of our calculation and comparison is done with totals. That is why we often need to focus on totals or the sum of all terms rather than on averages. Totals are much simpler to use in most calculations than averages. It's convenient to use the same formula for averages in a different form:

Sum of terms = Average · Number of terms

Let's name this form of the formula the "average formula for totals."

> **DEFINITION**
>
> The **arithmetic mean** is the most commonly used type of average and is often referred to simply as the *average*. The term *mean* (or *arithmetic mean*) is preferred in mathematics and statistics to distinguish it from other terms related to averages, such as the median and mode.

Average Problems

Let's solve several problems using the formula for arithmetic mean or average and see how it works.

Sample Problem 1

Find the average of $x - 1$, $x - 2$, $2x - 5$, $2x + 2$, and $1 - x$.

Step 1: The number of terms in the problem is 5. To find the average of these terms, we need to use the following formula for the average and substitute the known values:

$$\text{Average} = \frac{\text{Sum of Terms}}{\text{Number of Terms}}$$

Step 2: First, we need to add all five terms to obtain the sum of terms:

$(x - 1) + (x - 2) + (2x - 5) + (2x + 2) + (1 - x)$

Open parentheses and collect like terms:

$x - 1 + x - 2 + 2x - 5 + 2x + 2 + 1 - x = 5x - 5$

We found that the sum of all terms is $5x - 5$.

Step 3: Substitute the sum of all terms and the number of terms (5) into the formula in Step 1:

$$\text{Average} = \frac{5x - 5}{5}$$

Take out the *greatest common factor* (*GCF*) from the numerator and simplify:

DEFINITION

Greatest common factor (GCF) is the largest number that divides evenly into two or more numbers. For example, the GCF for 15 and 30 is 5.

$$\frac{5(x-1)}{5} = \frac{\cancel{5}(x-1)}{\cancel{5}} = x-1$$

We found that the average of five numbers is $x - 1$.

Solution: The average is $x - 1$.

Sample Problem 2

The average of a list of six numbers is 40. If we remove one number from the list, the average of the remaining five numbers is 30. What number was removed?

Step 1: Let x represent the removed number. The removed number could be obtained by finding the difference between the sum of the original 6 numbers and the sum of the remaining 5 numbers. We can state this mathematically as:

x = Sum of the original 6 numbers – Sum of the remaining 5 numbers

Step 2: We know that the average of the original numbers is 40 and that there are 6 of them. We calculate the sum of the original 6 numbers by using the formula for totals:

Sum of terms = Average · Number of terms

Sum of 6 original numbers = 40 · 6 = 240

We found that the sum of the original 6 numbers is 240.

Step 3: We know that the average of the remaining numbers is 30 and that there are 5 of them. We calculate the sum of the remaining numbers by using the same formula:

Sum of terms = Average · Number of terms

Sum of the remaining numbers = 30 · 5 = 150

We found that the sum of the remaining numbers is 150.

Step 4: Now we can calculate the removed number by substituting the values we found in Steps 2 and 3 into the equation from Step 1:

x = Sum of the original 6 numbers – Sum of the remaining 5 numbers

$x = 240 - 150 = 90$

Solution: The removed number is 90.

Now we apply the average formula to calculate the average score in class.

Sample Problem 3

Michael scored 84, 88, 77, and 87 on his first 4 tests. What score must he receive on the fifth test so that his average for all 5 tests is 85 or better?

Step 1: Let x represent the score for the fifth test. Let's also denote the sum of Michael's scores for the first 4 tests as S_4. We can calculate it by adding all 4 scores:

$S_4 = 84 + 88 + 77 + 87 = 336$

Step 2: When Michael receives his fifth score, his average score can be calculated using the following formula:

$$\text{Average} = \frac{\text{Sum of all Scores}}{\text{Number of Scores}}$$

The sum of all scores can be expressed as:

Sum of all scores = Sum of the first 4 scores + Fifth score

Sum of all scores = $S_4 + x$

Step 3: Use the totals form of the main formula to simplify the equation:

Sum of all scores = Average · Number of scores

$S_4 + x = 85 \cdot 5$

$S_4 + x = 425$

Isolate the variable:

$x = 425 - S_4$

Substitute the value for S_4 from Step 1:

$x = 425 - 336 = 89$

We found the score for Michael's fifth test.

Solution: Michael must get no less than 89 on his fifth test to sustain the average of 85.

Sample Problem 4

The Riviera Maya has 5 five-star all-inclusive resorts near the ancient city of Tulum. The first 3 have the following prices per night: $63, $84, and $96. The price per night for the fourth resort is $4 less than the price per night for the fifth resort. What are the prices per night for the last 2 resorts if the average price per night for all 5 resorts is $85?

Step 1: We don't know the prices for the last two resorts. Let the price for the fifth resort be x; then the price for the fourth resort is $x - 4$, since its price is $4 less.

Step 2: For this problem, we use the average formula for totals where we already know the average ($85) and the number of terms (5 resorts):

Sum of terms = Average · Number of terms = $85 \cdot 5 = 425$

We found that the total price per night for all 5 resorts is $425.

Step 3: Let's now express the sum of all terms by adding all the prices per night:

Sum of terms = $63 + 84 + 96 + x + (x - 4)$

Collect like terms:

Sum of terms = $2x + 239$

We found that the price per night for all 5 resorts is $2x + 239$.

A WORD OF ADVICE

Another type of average problem is average speed. We discuss these problems in Chapter 18.

Step 4: Substitute the last expression from Step 3 into the left side of the formula in Step 2:

Sum of terms = 425

$2x + 239 = 425$

Isolate the variable:

$2x = 186$

Divide both sides by 2:

$x = 93$

We found that the price per night for the last resort is $93.

Step 5: To find the price for the fourth resort, substitute $93 into the expression from Step 1:

$x - 4 = 93 - 4 = 89$

We found that the price for the fourth resort is $89.

Solution: The price for the fourth resort is $89 and for the fifth is $93.

Weighted Average Problems

Another type of average problems involves the weighted average, which is the average when all the values are not given equal weights.

The formula for the weighted average is:

$$\text{Weighted Average} = \frac{\text{Sum of Weighted Terms}}{\text{Total Number of Terms}}$$

To find the weighted terms, multiply each term by its weighting factor, which is the number of times each term occurs. Let's see how this formula works by doing several problems.

Sample Problem 5

A class of 25 students took a math test. Ten students had an average score of 75. Another part of the class had an average score of 65. What is the average score of the whole class?

Step 1: Since only 10 students received the score of 75, the number of students who received the lower score of 65 is 25 – 10 = 15 students.

Step 2: To obtain the sum of weighted terms, multiply each average (75 and 65) by the corresponding number of students who received that average (25 and 15, respectively) and then sum them up:

Sum of weighted terms = 75 · 10 + 65 · 15 = 750 + 975 = 1,725

We found that the sum of weighted terms is 1,725.

DEAD ENDS

You will get the wrong answer if you add the two average scores and divide the answer by 2.

Step 3: To find the weighted average, use the formula and plug in the sum of weighed terms that we found in Step 2 and the total number of terms, which is 25 (the number of students in the class):

$$\text{Weighted Average} = \frac{1,725}{25} = 69$$

We found that the average score is 69.

Solution: The average score of the whole class is 69.

Have you ever wondered why the average is called "weighted"? Does it have something to do with weights? Yes, it does. The number line can be interpreted as a weightless lever. The arithmetic mean (average) can be interpreted as the fulcrum on a number line where a set of identical weights are placed at different locations on the number line in such a way that the line is balanced. Let's illustrate this with an example. First, we calculate the average of four numbers—70, 80, 90, and 100—using the formula for average:

$$\text{Average} = \frac{70+80+90+100}{4} = 85$$

Then we place these numbers on the number line with the fulcrum at the number 85 and attach 1-unit weights to each number.

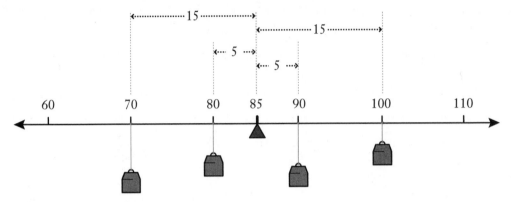

If we assume that the distance between every two integers on the number line is 1 length unit, then the distance between the number 70 and the fulcrum is 15 (85 – 70), between the number 90 and the fulcrum is 5 (90 – 85), and so forth.

Write the lever equation using the fact that every number in the set carries the same 1-unit weight and using the distances from each number to the fulcrum:

$w_1d_1 + w_2d_2 = w_3d_3 + w_4d_4$

$(1)(15) + (1)(5) = (1)(5) + (1)(15)$

$15 + 5 = 5 + 15$

$20 = 20$

Both sides are equal, so that means the lever is balanced.

Since the lever is balanced, we can conclude that the average 85 of the set of numbers 70, 80, 90, and 100 is the fulcrum of the lever (number line).

Now let's discuss what happens if the weights that we attach to the numbers are different and one number in the set will carry more weight. How will it influence the average? Let's attach the same 1-unit weight to the numbers 70, 80, and 90. Then we attach two 1-unit weights to the number 100.

First, let's calculate the weighted average using the formula:

$$\text{Weighted Average} = \frac{1 \cdot 70 + 1 \cdot 80 + 1 \cdot 90 + 2 \cdot 100}{5} = \frac{440}{5} = 88$$

Note that we calculated the number 100 two times because it has two weights.

We found that the weighted average of the set of numbers 70, 80, 90, and 100 is 88.

Now let's prove with the lever formula that 88 is indeed the weighted average for our set of numbers.

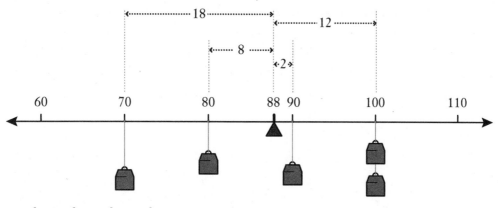

$$w_1 d_1 + w_2 d_2 = w_3 d_3 + w_4 d_4$$

Substitute the weights (don't forget to plug in 2 as the weight for the number 100) and the corresponding distances:

$$(1)(18) + (1)(8) = (1)(2) + (2)(12)$$

$$18 + 8 = 2 + 24$$

$$26 = 26$$

Both sides are equal, so the lever is balanced. Since the lever is balanced, we can conclude that the weighted average 88 of the set of numbers 70, 80, 90, and 100 is the fulcrum of the lever.

By doing these exercises with a lever and averages, we gave the arithmetic mean a physical meaning as the fulcrum on the number line.

The concept of a weighted average is a common grading model used in schools and colleges. Different types of work performed by the students are assigned a value (weight) that helps calculate the final grade. Homework may account for a smaller percentage of the total grade, while several tests may carry additional weight in the final grade, along with a group project and a final exam. This means that tests and the final exam carry more importance in securing a good grade, although the successful performance on other components ensures an even higher grade.

Let's do the next problem and see how teachers calculate our final grade.

Sample Problem 6

Find the average score that Tom has in his science class using the following table, where the type of assignment is recorded in the first column, the weight for each assignment is recorded in the second column, and Tom's corresponding scores are listed in the third column:

Assignment	Weight	Tom's Score
Homework	1	92
Test 1	2	100
Test 2	2	90
Test 3	2	80
Group project	3	90
Final exam	4	100

Step 1: To simplify our calculations, let's expand the table and add a column for the value of each assignment and add a row for the total number of weights and the total value. The value for each assignment is calculated by multiplying Tom's score by the weight of each assignment. The total for weight is obtained by adding all the weights together. Finally, the total for value is obtained by adding the values for all assignments:

Assignment	Weight	Tom's Score	Value
Homework	1	92	92
Test 1	2	100	200
Test 2	2	90	180
Test 3	2	80	160
Group project	3	90	270
Final exam	4	100	400
Totals	14		1,302

A WORD OF ADVICE

Creating a table for weighted average problems helps to simplify calculations and avoid mistakes.

Step 2: To calculate Tom's average score for the science class, use the formula for the weighted average and substitute the sum of weighted terms (total value) and the total number of terms (total number of weights) from Step 1:

$$\text{Weighted Average} = \frac{\text{Sum of Weighted Terms}}{\text{Total Number of Terms}}$$

$$\text{Weighted Average} = \frac{1,302}{14} = 93$$

We found that Tom's average score is 93.

Solution: Tom's average score for the science class is 93.

In general, you can assign any weights, not necessarily integers. Often when we choose the fractional weights, they add up to 1.

Sample Problem 7

Find the average score that Sara has for her midterm in her English class using the following table, where the type of assignment is recorded in the first column, the weight for each assignment is recorded in the second column, and Sara's corresponding scores are listed in the third column:

Assignment	Weight	Sara's Score
Homework	$\frac{1}{12}$	72
Test	$\frac{1}{4}$	80
Group project	$\frac{1}{6}$	90
Midterm exam	$\frac{1}{2}$	100

Step 1: Let's again expand the table and add a column for the value of each assignment and add a row for the total number of weights and the total value. The value for each assignment is calculated by multiplying Sara's score by the weight of each assignment. For example, the value for the homework can be calculated as:

$$\frac{1}{12} \cdot 72 = \frac{72}{12} = 6$$

The total for weights is obtained by adding all the weights:

$$\frac{1}{12} + \frac{1}{4} + \frac{1}{6} + \frac{1}{2}$$

Using 12 as the least common denominator (LCD), convert all fractions into the equivalent fractions with the denominator of 12 and add them:

$$\frac{1}{12} + \frac{3}{12} + \frac{2}{12} + \frac{6}{12} = \frac{12}{12} = 1$$

Note that the total weight for all assignments conveniently adds up to 1.

Finally, the total for values is obtained by adding the values for all assignments:

Assignment	Weight	Sara's Score	Value
Homework	$\frac{1}{12}$	72	6
Test	$\frac{1}{4}$	80	20
Group project	$\frac{1}{6}$	90	15
Midterm exam	$\frac{1}{2}$	100	50
Totals	1		91

A WORD OF ADVICE

When weights for each assignment are given as fractions or percents, the total weight often adds up to 1 or 100% (1 in decimal form).

Step 2: To calculate Sara's average score for the midterm in her English class, use the formula for the weighted average and substitute the sum of weighted terms (total value) and the total number of terms (total number of weights) from Step 1:

$$\text{Weighted Average} = \frac{\text{Sum of Weighted Terms}}{\text{Total Number of Terms}}$$

$$\text{Weighted Average} = \frac{91}{1} = 91$$

We found that Sara's average midterm score is 91.

Solution: Sara's average score for the midterm in her English class is 91.

The last problem for this chapter concerns grade point average (GPA). Most colleges assign "weights" to the individual course grades in the form of credits. An A grade in a 4-credit course affects your GPA 50% more than an A grade in a 2-credit course. Most colleges use this scale: A = 4, B = 3, C = 2, D = 1, F = 0.

WORTHY TO KNOW

In addition to the average scores and GPA problems, you can solve many other problems using the weighted average formula. For example, the formula helps calculate the average price per pound for several types of cookies if you know the unit price per each type and the amount in pounds for each that has been sold.

Sample Problem 8

John took the following courses and received the following grades:

Course	Credits	Grades
Calculus	4	C
English Literature	2	A
Computer Science	3	B
Psychology	2	B

Calculate John's GPA.

Step 1: Let's expand the table and add a column for the value of each course and add a row for the total number of credits and the total course value. The value for each course is calculated by multiplying the course credit by John's grade (converted to the number equivalent). The total number of credits is obtained by adding all credits. The total course value is calculated by adding values for all courses:

Course	Credits	Grades	Values
Calculus	4	2	8
English Literature	2	4	8
Computer Science	3	3	9
Psychology	2	3	6
Total	11		31

Step 2: To calculate John's GPA, use the formula for the weighted average and substitute the total values of all courses and the total number of credits from Step 1:

$$\text{Weighted Average} = \frac{\text{Sum of Weighted Terms}}{\text{Total Number of Terms}}$$

$$\text{Weighted Average} = \frac{31}{11} = 2.818 \approx 2.82$$

We found that John's GPA is 2.82.

Solution: John's GPA is 2.82.

Practice Problems

Problem 1: Sam's average score for his first 3 tests is 83. What must he get on his fourth test so that his average score for the 4 tests is no less than 86?

Problem 2: Find the average of $x - 3$, $2x - 4$, $2x - 2$, $1 - x$, x, and $x - 4$.

Problem 3: The average of a list of 5 numbers is 30. If we add 1 number to the list, the average of the 6 numbers will be 32. What number was added?

Problem 4: Sandy bought 5 types of chocolate candies for the party. The prices per pound for the first 3 types were $5, $7, and $8. The price for the fourth type is $3 less than for the fifth type. Find the price per pound for the fourth and the fifth types of candies if the average price per pound for all 5 types was $7.

Problem 5: A Tea for All Tastes shop sells different types of tea from around the world. Last week, they sold 100 pounds of Indian tea for $8 per pound, 60 pounds of Ceylon tea for $9 per pound, 40 pounds of Russian tea for $7 per pound, and 50 pounds of Chinese tea for $8 per pound. What was the average cost per pound of the tea sold last week?

Problem 6: In Ms. Peterson's English class, homework counts 10%, quizzes 20%, group projects 30%, and tests 40%. If Morris has a homework grade of 92, a quiz grade of 68, a group projects grade of 75, and a test grade of 94, what is his average grade for the class?

Problem 7: Maria received an A for her 4-credit algebra class, a B for her 3-credit history class, a C for her 3-credit English class, and an A for her 2-credit music class. Compute her GPA.

The Least You Need to Know

- The main formula for average problems is: Average $= \dfrac{\text{Sum of Terms}}{\text{Number of Terms}}$.

- It is often convenient to use the average formula for totals (sum of terms): Sum of terms = Average · Number of terms.

- The formula for weighted average is:

  Weighted Average $= \dfrac{\text{Sum of Weighted Terms}}{\text{Total Number of Terms}}$.

- The concept of a weighted average is a common grading model used in schools and colleges for computing grades and GPA.

Motion Problems with a Single Traveler

In This Chapter

- Breaking motion problems into parts
- Choosing the right variable
- Making the units uniform
- Visualizing the problem with a diagram

"Motion" or "distance" word problems involve something or somebody traveling at some steady speed (rate). To recognize a motion or distance problem, the key phrases to look for are *how fast*, *how far*, or *how long*. In this chapter, we look at motion problems with a single traveler, learn to calculate the average speed for the entire trip, and tackle round-trip problems.

Motion Problems Basics

As we've seen, each category of word problem shares common elements. In motion problems, you're asked to find the value of one of three variables: distance, time (t), or speed/rate (r). The main formula for all motion problems is:

$d = r \cdot t$

d stands for distance.

r stands for the (constant or average) speed or rate.

t stands for time.

Calculating Distance

When the rate and time are known, we use the main formula directly to find the distance traveled:

$d = r \cdot t$

For example, if a car travels at an average speed of 60 mph, how far will it travel in 3 hours? We know the speed and time, and we can plug them into the main formula to find the distance:

$d = 60 \text{ mph} \cdot 3 \text{ h} = 180 \text{ miles}$

Solution: The car traveled 180 miles.

Calculating Speed

When the distance and time are known and we need to find the speed or rate, remember that you can calculate the rate/speed by dividing the distance gone by the time required to go that distance:

$r = \dfrac{d}{t}$

For example, the distance between two towns is 40 miles, and a bicyclist traveled for 4 hours. What was his average speed during his trip? We know the distance and time, so we can find the speed by substituting the values for distance and time into the previous formula:

$r = \dfrac{40 \text{ miles}}{4 \text{ h}} = 10 \text{ mph}$

Solution: The average speed of the bicyclist was 10 mph.

Calculating Time

When the distance and speed are given and we need to find the time, we use the formula in this form because the time required is the result of the distance divided by the rate/speed:

$t = \dfrac{d}{r}$

For example: A camper leaves a campsite to walk to another camp that is 15 miles away. His walking rate is 3 mph. How long will it take for him to reach the second camp? In this case, the distance and average speed are known, and we need to find the time. Let's use the formula:

$$t = \frac{15 \text{ miles}}{3 \text{ mph}} = 5 \text{ h}$$

Solution: It will take him 5 hours to reach the second camp.

A WORD OF ADVICE

Because these problems usually involve a story of a journey, it is helpful to create a sketch of the trip. A sketch helps translate the words of the problem into an actual portrait, to sort out the variables more easily.

Single Traveler, Single Direction

We already know how to find an average speed when the distance and time are given. But what if someone is moving with one average speed for the first part of a trip, and then increases or decreases that average speed for the second part of the trip? This affects the average speed of the entire trip. Let's do several problems and discuss this.

Sample Problem 1

A grandma traveled 80 miles to visit her grandchildren. For the first half of the trip distance, she averaged 60 mph, and for the second half, 40 mph. What was the average speed for the total distance traveled?

At the beginning of this chapter, we learned that if the distance and time are known and we need to find the speed or rate, we use the formula in the following form:

$$r = \frac{d}{t}$$

In our case, we don't know the total time it took the grandma to reach her grandchildren, so we can't use the formula right away. We need some additional steps to find the time and then use the formula.

Step 1: Let t_1 represent the time for the first part of travel and t_2 represent the time for the second part of travel.

Step 2: Using the variables from Step 1, sketch a picture of the problem situation.

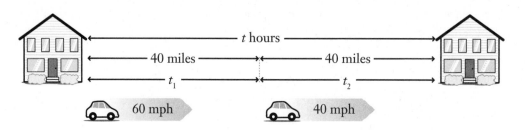

Step 3: Looking at the sketch, we can state that the overall time of travel is the sum of the two times: the time of the first part of travel and the time of the second part of travel. In algebra language, this statement will be:

$t = t_1 + t_2$

Step 4: Now we need to find the values for t_1 and t_2. For the first part of the trip, we know the distance and the average speed, so we can find the time using the simple time calculation form we explained earlier:

$$t_1 = \frac{40 \text{ miles}}{60 \text{ mph}} = \frac{2}{3} \text{ h}$$

For the second part of the trip, we also know the distance and the average speed, so we can find the time:

$$t_2 = \frac{40 \text{ miles}}{40 \text{ mph}} = 1 \text{ h}$$

Let's plug t_1 and t_2 into the formula in Step 3 to find t:

$$t = \frac{2}{3} + 1 = 1\frac{2}{3}$$

Grandma traveled a total of $1\frac{2}{3}$ hours.

Step 5: We found the overall time of the trip. Going back to the general formula for calculating speed, we can calculate the average speed for the whole trip:

$$r = \frac{d}{t} = d \div t = 80 \div 1\frac{2}{3}$$

Note that $1\frac{2}{3}$ hours translate into $\frac{5}{3}$ hours in the formula:

$$\frac{80}{1} \div \frac{5}{3} = \frac{80}{1} \cdot \frac{3}{5} = \frac{240}{5} = 48 \text{ mph}$$

Solution: The average speed for the entire trip was 48 mph.

DEAD ENDS

Make sure you're using the same units for time, distance, and rate in each problem. For instance, if you're given a rate in kilometers per hour, then time must be in hours (not in minutes or seconds) and distance must be in kilometers (not in miles). Don't allow the problem to trick you by using the wrong units. If the units in the problem are different, always convert them to the correct ones so that all units are uniform.

Sample Problem 2

Sam has an old car that always gives him trouble when he least expects it. Last week he drove home from college. For 48 minutes, Sam traveled at a certain speed. Then car trouble forced him to reduce his speed by 30 miles for the remainder of the trip. If the total distance Sam traveled was 85.5 miles, and it took him 2 hours and 3 minutes to finish the entire trip, how far did he drive at the slow speed?

To find the distance Sam traveled at the slow speed, we need to find the speed and time for the second part of travel.

Step 1: First, we assign variables. Let Sam's speed for the first part of his trip be r. Since he reduced his speed by 30 mph, the speed for the second part will be $r - 30$.

Step 2: Let's also represent total distance traveled (85.5 miles) as d, the first half of Sam's journey as d_1, and the second half as d_2.

Step 3: Make a sketch of the problem situation using the variables from Steps 1 and 2. We indicate the point where Sam had to reduce his speed with the letter C on the diagram.

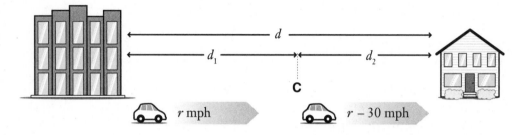

Step 4: The next factor to consider is time traveled. We already know how long the entire trip took (2 hours and 3 minutes). We also know how long Sam traveled at the higher speed (48 minutes). Now we need to find how long he traveled at the slower speed by subtracting the first time period (48 minutes) from the total time traveled (2 hours and 3 minutes). Since we need to subtract like units, convert the total time into minutes:

2 hours + 3 minutes = 2(60 min) + 3 min = 120 min + 3 = 123 min

Then subtract 48 minutes:

123 – 48 = 75 minutes

We found that the time traveled at the slower speed was 75 minutes.

Since speed is calculated by miles per hour, we need to convert the minutes into hours. Forty-eight minutes becomes $\frac{48}{60} = \frac{4}{5}$ h (reduced by the common factor of 12), and 75 minutes becomes $\frac{75}{60} = \frac{5}{4}$ h (reduced by the common factor of 15). We found that Sam traveled at the slow speed $\frac{5}{4}$ hours.

Step 5: It is obvious from the sketch that the total distance is the sum of the first and second distances:

$d = d_1 + d_2$

To solve the problem, we have to find out how far Sam drove at each speed; then we can calculate the actual speed he traveled. We can express the first distance using the main distance formula:

$d_1 = r \cdot \frac{4}{5} = \frac{4r}{5}$

r is the speed for the first part, and $\frac{4}{5}$ h (48 minutes) is the time it took Sam to drive this distance.

Similarly, we can express the part of the distance driven at the reduced speed as:

$$d_2 = (r-30)\cdot\frac{5}{4} = \frac{5(r-30)}{4}$$

$(r - 30)$ is the speed for this part of travel, and $\frac{5}{4}$ h (75 minutes) is the time.

Step 6: Now plugging the actual value for d (85.5 miles) and the two expressions for d_1 and d_2 into the first formula in Step 5 gives us the equation that will help us find the rate at which Sam traveled before his car broke down:

$$\frac{4r}{5} + \frac{5(r-30)}{4} = 85.5$$

To eliminate fractions, let's multiply each term of the equation by the least common denominator (LCD), which is 20.

$$\frac{20\cdot 4r}{5} + \frac{20\cdot 5(r-30)}{4} = 20\cdot 85.5$$

Reduce each term:

$$\frac{\overset{4}{\cancel{20}}\cdot 4r}{\underset{1}{\cancel{5}}} + \frac{\overset{5}{\cancel{20}}\cdot 5(r-30)}{\underset{1}{\cancel{4}}} = 20\cdot 85.5$$

Multiply the remaining factors:

$16r + 25(r - 30) = 1{,}710$

Distribute the left side:

$16r + 25r - 750 = 1{,}710$

Combine like terms and isolate variable:

$41r = 2{,}460$

Divide both sides by 41:

$r = 60$ mph

Step 7: We found that the average speed for the first part of the trip, before car trouble, was 60 mph. We can now calculate the reduced speed for the second part of travel:

$r - 30 = 60 - 30 = 30$ mph

We found that Sam's speed for the second part of travel was 30 mph.

Knowing the speed and time, we can finally answer the problem's question of "How far did he drive at the slow speed?"

$$d_2 = 30 \text{ mph} \cdot \frac{5}{4} \text{ h} = \frac{30 \cdot 5}{4} = 37.5 \text{ miles}$$

Solution: Sam drove at the slow speed for 37.5 miles.

> **A WORD OF ADVICE**
>
> When you need to convert minutes into hours, simply write the fraction that has minutes as its numerator and 60 as the fraction's denominator, and then reduce the fraction, if possible. For instance, to convert 15 minutes into hours, write $\frac{15}{60} = \frac{1}{4}$ h.

The Round-Trip

In this type of motion problem, the common theme is that a person or object makes a round-trip. And since on their way back they travel the same distance as they travel on their way to the destination, we equate these distances.

Sample Problem 3

James drove to a nearby town at an average speed of 40 mph. The next day, he drove back at an average speed of 30 mph. If he spent a total of 7 hours traveling, what is the distance James traveled?

Step 1: Remember that the distance James traveled to the nearby town is the same that he traveled back home. Let's represent this distance as d.

Step 2: The total time of travel t is the sum of the time it took James to reach the town t_1 and the time he traveled back home t_2.

$$t_1 + t_2 = t$$

Step 3: For the first part of the trip, we know James's speed (40 mph), so we can express t_1 using the simple time calculation we discussed at the beginning of the chapter as:

$$t_1 = \frac{d}{40}$$

For the second part of the trip back home, we also know his speed (30 mph), so we can express t_2 as:

$$t_2 = \frac{d}{30}$$

Step 4: Plugging the two expressions for t_1 and t_2 from Step 3 into the formula in Step 2 and using the actual value for t (7 hours), we obtain the following equation:

$$\frac{d}{40} + \frac{d}{30} = 7$$

Multiply each term by the LCD, which is 120:

$$\frac{120d}{40} + \frac{120d}{30} = 120 \cdot 7$$

Reduce each term by the common factor:

$$\frac{\overset{3}{\cancel{120}}d}{\underset{1}{\cancel{40}}} + \frac{\overset{4}{\cancel{120}}d}{\underset{1}{\cancel{30}}} = 120 \cdot 70$$

Multiply the remaining factors:

$$3d + 4d = 840$$

Collect like terms and divide both sides by 7:

$$7d = 840$$

$$d = 120 \text{ miles}$$

Step 5: We found that the distance between the towns is 120 miles. The total distance James traveled will be twice that distance:

$$2 \cdot 120 \text{ miles} = 240 \text{ miles}$$

Solution: The total distance James traveled is 240 miles.

Practice Problems

Problem 1: What is the distance a plane can travel from 1:30 A.M. to 7:00 A.M. flying at a rate of 130 mph?

Problem 2: What is the average speed of a jogger who traveled 15 miles in 2 hours and 30 minutes?

Problem 3: On a 60-mile bicycle trip, Andrea rode 40 miles at 20 mph and the rest of the trip at 10 mph. What was Andrea's average speed on the bike trip?

Problem 4: A girl runs at a rate of 5 mph for 2 hours and then rides her bike at a rate of 10 mph for 3 hours. What is her average rate for the entire trip?

Problem 5: George rode out into a rural area at a speed of 30 mph and back home at a speed of 35 mph. His round-trip took 6.5 hours. How far out did he go?

Problem 6: A scientist climbed up a mountain to an observation post at an average speed of 2 mph. He conducted some experiments for 2 hours and returned back to the bottom of the mountain at an average speed of 4 mph. The entire trip took 8 hours. What is the distance to the observation post from the bottom of the mountain?

Problem 7: Marilyn travels 5 hours to and from work every day. Her average rate of speed to work is 9 mph, and her average rate of speed returning is 6 mph. How far is it from her home to work?

The Least You Need to Know

- The basis of all distance problems is the formula $d = r \cdot t$.
- Rate is the same as speed.
- When you encounter a motion problem, first identify its type from the wording and given facts, then assign variables.
- In the round-trip problems, the distance to the destination is the same as the distance back to the starting point.

Motion Problems with Multiple Travelers

Chapter

19

In This Chapter

- Summing up the distances
- Dealing with time in motion problems
- Exploring the effect of wind and current
- Making a chart using the main distance formula

In this chapter, we explore motion problems when two parties are traveling. It can be two cars traveling toward each other, or two people walking in the opposite direction, or one train catching up with another. It can even be the wind or a stream moving at a certain speed whose movement affects the speed of an object moving in this medium.

Moving Toward Each Other

In this type of problem, it is important to remember that, regardless of the speeds or the distance traveled, the parties will take the same amount of time to meet.

Sample Problem 1

Two towns are located 460 miles apart. A car leaves the first town traveling toward the second town at 60 mph. At the same time, a motorcyclist leaves the other town and heads toward the first town at 55 mph. How long will it take for the drivers to meet?

Step 1: Since both drivers started traveling at the same time and meet at the same time somewhere in between, it took the car and the cyclist exactly the same time to meet. Let t indicate this time. Since the rates are different, the distances they both traveled are different as well: the car traveled the longer distance, since its rate (speed) is slightly higher.

Step 2: Sketch a picture, where the letter C represents the meeting point:

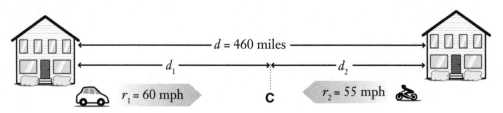

Step 3: Looking at this picture, we can see that the distance between the two towns equals the sum of the two distances: the distance traveled by the car (d_1) and the distance traveled by the cyclist (d_2). In algebra language, this statement is:

$d_1 + d_2 = d$

Step 4: Now we can express distances d_1 and d_2 using the simple distance formula:

$d_1 = r_1 t$

$d_2 = r_2 t$

r_1 and r_2 are the rates of the car and the cyclist, respectively, and t is time each driver traveled before meeting.

Step 5: Plugging the two expressions for d_1 and d_2 from Step 4 into the formula in Step 3 gives us the equation that will help us find the time it took both drivers to meet:

$r_1 t + r_2 t = d$

The values for r_1, r_2, and d are given, and we can plug them into the preceding equation:

$60t + 55t = 460$

To solve this equation, we combine like terms and divide both sides of the equation by 115:

$115t = 460$

$t = 4$ hours

Solution: It will take the car and the motorcyclist 4 hours to meet.

> **A WORD OF ADVICE**
>
> Usually there's more than one way to solve a motion problem. You have the freedom to choose the variable, although it's best to choose the one that allows you to find the answer in the shortest and easiest way.

Sample Problem 2

Two airports are 950 miles apart. One plane leaves from each airport headed for the other: a twin-propeller and a jet. In 30 minutes, the distance between them is 150 miles. What is the speed of the jet if it flies 3 times faster than the twin-propeller-powered aircraft?

Step 1: The first factor to consider is speed. Let's say that the speed of the twin-propeller is r. Then the speed of the jet is $3r$, since it is 3 times faster.

Step 2: Let's now deal with the distance factor. We know that the distance between the airports is 950 miles. After flying 30 minutes, the pilots decreased that distance to 150 miles. That means that, in 30 minutes, both aircraft traveled the distance:

950 – 150 = 800 miles

Since speed is measured in miles per hour, let's convert 30 minutes into hours:
$\dfrac{30}{60} = \dfrac{1}{2}h$.

Step 3: The small aircraft flew the distance that we denote as d_1, and the jet traveled the distance that we denote as d_2. We know that the combined distance traveled by both aircraft is 800 miles. We can express this fact mathematically as:

$d_1 + d_2 = 800$

Step 4: Using the variables and values from Steps 1, 2, and 3, we can sketch a picture of the problem.

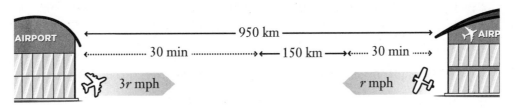

Step 5: The distance the twin-propeller traveled in $\frac{1}{2}h$ (30 minutes) at the speed of

r mph can be expressed using the simple calculation for the distance as:

$$d_1 = r \cdot \frac{1}{2} = \frac{r}{2}$$

The distance the jet traveled in $\frac{1}{2}h$ at the speed of $3r$ mph can be expressed as:

$$d_2 = 3r \cdot \frac{1}{2} = \frac{3r}{2}$$

Step 6: Plugging the two expressions for distances d_1 and d_2 into the formula in Step 3, we obtain the equation that will help us find the speed of the twin-propeller:

$$\frac{r}{2} + \frac{3r}{2} = 800$$

To eliminate fractions, multiply both parts of the equation by the least common denominator (LCD), which is 2:

$$\frac{2 \cdot r}{2} + \frac{2 \cdot 3r}{2} = 2 \cdot 800$$

Reduce by the common factors:

$$\frac{\overset{1}{\cancel{2}}r}{\underset{1}{\cancel{2}}} + \frac{\overset{1}{\cancel{2}} \cdot 3r}{\underset{1}{\cancel{2}}} = 2 \cdot 800$$

Multiply the remaining factors:

$r + 3r = 1{,}600$

$4r = 1{,}600$

$r = 400$ mph

DEAD ENDS

If two parties started to move either toward each other or in opposite directions at the same time, don't introduce two times—the time of travel is the same for both parties.

Step 7: We found the speed of the twin-propeller. Since the speed of the jet is 3 times the speed of the twin-propeller, we multiply the twin-propeller's speed by 3:

$3r = 3(400) = 1{,}200$ mph

Solution: The speed of the jet is 1,200 mph.

Moving in Opposite Directions

This type of motion problem shares two key elements with problems 1 and 2: both parties start at the same time and spend the same amount of time traveling. The difference is simply that they don't meet.

Sample Problem 3

Ashley and Bob went on separate business trips. They started from their house and traveled in opposite directions. Ashley averaged 55 mph, while Bob averaged 65 mph. In how many hours were they 360 miles apart, when each of them stopped for lunch and to talk to each other on the phone for an hour?

Step 1: Let's say that the time that each of them traveled before they stopped for lunch is t. In this time, the distance they traveled together is 360 miles.

Step 2: The total distance they traveled together is 360 miles. We denote the distance that Ashley traveled as d_1 and the distance that Bob traveled as d_2. Then the total distance can be expressed as:

$d_1 + d_2 = 360$

Step 3: Using the variables from Steps 1 and 2, sketch a picture about the problem situation:

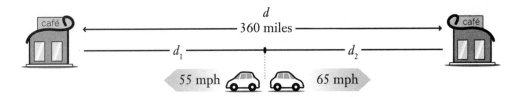

Step 4: The distance d_1 that Ashley traveled can be expressed using the simple distance formula and Ashley's actual speed (55 mph):

$d_1 = 55t$

Similarly, the distance d_2 that Bob traveled can be written as:

$d_2 = 65t$

Step 5: By plugging the two expressions for d_1 and d_2 into the formula in Step 2, we obtain the main equation for the problem:

$55t + 65t = 360$

Combine like terms:

$120t = 360$

Divide both sides by 120:

$t = 3$ hours

Solution: Ashley and Bob will be 360 miles apart in 3 hours.

One piece of information we never used while solving the problem: the fact that Ashley and Bob talked to each other on the phone for an hour. As we've seen, this information is irrelevant to the problem-solving process and can be ignored.

Moving in the Same Direction

In this type of motion problem, the key factors are departure time and traveling speed. Another important point to notice is that both objects are covering exactly the same distance in different amounts of time.

Sample Problem 4

A train traveling at a speed of 96 mph is to overtake another train that has a head start of 2 hours and is traveling at a speed of 72 mph. How long will it take the faster train to overtake the other train?

Step 1: As we mentioned before, both trains will travel exactly the same distance before the faster one will overtake the slower train. This is important to understand. The equation that states this fact mathematically is:

$d_1 = d_2$

d_1 and d_2 are the distances traveled by the fast train and the slow train, respectively.

A WORD OF ADVICE

In catching-up problems, equating the distances of both parties helps to write the problem's equation.

Step 2: Now we can deal with the time factor. Let's denote the time that the faster train took to overtake the slower train as t. Then the time that the slower train traveled before the faster train catches it is $t + 2$, since the slower train had a head start of 2 hours.

Step 3: Let's do our sketch, with the letter C indicating the catching point:

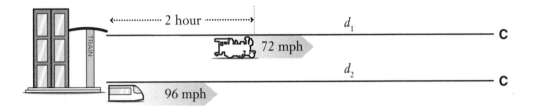

Step 4: Knowing the actual speeds of both trains (96 mph for the faster train and 72 mph for the slower train) and the time of travel for both trains from Step 2, we can express the distances d_1 and d_2 using the simple distance formula as:

$d_1 = 96t$

$d_2 = 72(t + 2)$

Step 5: By plugging the two expressions for d_1 and d_2 into the formula in Step 1, we obtain the equation that will help us find the time:

$96t = 72(t + 2)$

Distribute the right side of the equation and isolate the variable:

$96t = 72t + 144$

$24t = 144$

Divide both sides by 24:

$t = 6$ hours

Solution: It will take the faster train 6 hours to overtake the slower train.

Sample Problem 5

A boy on the way to visit his girlfriend learned that he had just missed his bus and that the next bus would leave in 38 minutes. He was a good runner and decided to run to his girlfriend's house instead. The boy ran at an average speed of 12 mph and reached her home at the same time as the next bus. If the bus traveled at an average speed of 50 mph, how far was the boy from his girlfriend's house when he missed the first bus?

Let's first point out that the catching point was the girlfriend's house, because both the boy and the next bus arrived there at the same time. So, really, the boy didn't save any time; he just preferred to run and enjoy the weather than wait at the bus stop.

Step 1: Going back to the problem, we can state that the distances that both the boy and the bus traveled were exactly the same. Let's denote this distance as d. Let's also convert the minutes into hours: $38 \min = \dfrac{38}{60} = \dfrac{19}{30} b$.

Step 2: Using the variables and values from Step 1, sketch a picture of the problem situation:

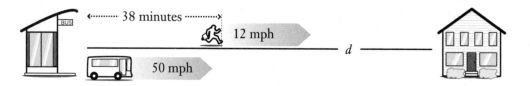

Step 3: We can express the boy's running time using the simple time formula and plug in the boy's actual speed (12 mph):

$$t_1 = \frac{d}{12}$$

Similarly, using the bus's actual speed (50 mph), the bus's time can be expressed as $t_2 = \dfrac{d}{50}$.

Step 4: To write the equation, we use the fact that the boy had a 38-minute ($\dfrac{19}{30}$ hours) head start. This means that, to equalize their times, we need to add these 38 minutes to the bus's time, since the bus started the trip later:

$$t_1 = t_2 + \frac{19}{30}$$

> **A WORD OF ADVICE**
>
> Sometimes in catching-up problems it is more convenient to write the problem's equation about the time of travel.

Step 5: Plugging the two expressions for t_1 and t_2 from Step 3 into the formula in Step 4, we obtain the equation that helps us find the distance:

$$\frac{d}{12} = \frac{d}{50} + \frac{19}{30}$$

Multiply each term of the equation by the LCD, which is 300:

$$\frac{300d}{12} = \frac{300d}{50} + \frac{300 \cdot 19}{30}$$

> **WORTHY TO KNOW**
>
> During ancient times, Zeno of Elea created so-called paradoxes of motion to support his teacher Parmenides' doctrine that motion is nothing but an illusion.

Reduce by the common factors to eliminate fractions:

$$\frac{\overset{25}{\cancel{300}}d}{\underset{1}{\cancel{12}}} = \frac{\overset{6}{\cancel{300}}d}{\underset{1}{\cancel{50}}} + \frac{\overset{10}{\cancel{300}} \cdot 19}{\underset{1}{\cancel{30}}}$$

Multiply the remaining factors:

$25d = 6d + 190$

Isolate the variable:

$19d = 190$

Divide both sides by 19:

$d = 10$ miles

Solution: The boy was 10 miles from his girlfriend's house.

With or Against the Elements

What makes this category of problem unique is the introduction of an element, usually wind or current, that affects the speed of the person or vehicle traveling. When an object is moving in the direction of the wind or current flow, the object's speed becomes the sum of its speed in still air or water plus the speed of the wind or current. When the object is moving against the wind or current, the resulting object's speed is the difference between the object's real speed and the speed of the wind or current. Another important point to note is that the object is usually traveling exactly the same distance with or against the wind or current.

Sample Problem 6

Julie rowed her boat from her vacation house upstream to her friends' house in 5 hours. She spent the evening and the next morning with her friends, and left their home after lunch. Julie rowed back to her house in 3 hours. If the stream is flowing at 2 mph, how fast can Julie row in still water?

Step 1: Let's start by denoting Julie's speed in still water as r. Then her speed upstream decreases by 2 mph (the speed of the stream) $(r-2)$, and her speed downstream increases by 2 mph $(r+2)$.

A WORD OF ADVICE

Creating a chart instead of a diagram using the main distance formula is useful in solving motion problems that involve traveling with or against the wind or current.

Step 2: Now we can set up a chart using the main distance formula $d = r \cdot t$. In this chart, the second column lists the speeds we found in Step 1 and the third column contains times for upstream and downstream travel. In the fourth column, we calculate the corresponding distances by multiplying speeds from the second column by the times from the third one.

	r	t	d
Upstream	$r-2$	5	$5(r-2)$
Downstream	$r+2$	3	$3(r+2)$

Step 3: Since Julie traveled the same distance both ways, we can express this fact mathematically by equalizing expressions for distance from column four:

$5(r-2) = 3(r+2)$

Distribute both sides:

$5r - 10 = 3r + 6$

Isolate variable and collect like terms:

$2r = 16$

Divide both sides by 2:

$r = 8$ mph

Solution: Julie's speed in still water is 8 mph.

Sample Problem 7

A helicopter flying with a tailwind (with the wind) flew 1,320 miles between 2 cities in 5 hours and 30 minutes. It took the helicopter 6 hours to fly the same distance against the wind. Find the speed of the helicopter in still air and the speed of the wind.

This problem looks similar to the previous one. However, don't be fooled. In the first problem, we knew the stream speed. Second, we were asked to find the value of one variable. In this problem, we are asked to find the speed of the wind and the helicopter in still air. As a result, we need to use the system of two linear equations to solve the problem.

Step 1: Let's denote the speed of the helicopter in still air as h and the wind speed as w. Then, when flying with the wind, the helicopter's speed increases by the wind speed and is ($h + w$); when flying against the wind, its speed decreases by the wind speed and is ($h - w$).

Step 2: Let's set up a chart using the main distance formula $d = r \cdot t$. The second column lists the speeds we found in Step 1, and the third column is for the times given by the problem. In the fourth column, we multiply expressions from the second and third columns to obtain the corresponding distances. Be sure to convert 5 hours and 30 minutes into hours:

5 h + 30 min. = 5.5 h

	r	t	d
Tailwind	$h + w$	5.5	$5.5(h + w)$
Headwind	$h - w$	6	$6(h - w)$

The two expressions in the fourth column are equal (the helicopter traveled the same distance with or against the wind). However, equalizing these distances as we did in the previous problem doesn't help, since our equation will have two unknown variables h and w, and we won't be able to find them using just one equation. In the case of two unknown variables, we need to set up a system of two linear equations.

Step 3: Before we set up the system of two linear equations, let's write these equations separately and simplify them. Since we know the actual distance (1,320 miles), we can write the equation for traveling with the tailwind by equalizing this value to the corresponding expression for the distance from the fourth column:

$5.5(h + w) = 1,320$

Dividing both sides by 5.5, we obtain the first simple equation:

$h + w = 240$

Similarly, we can write the equation for the travel with the headwind as:

$6(h - w) = 1,320$

After dividing both sides by 6, we obtain the second simple equation:

$h - w = 220$

We obtained two simple equations with two unknown variables. Now we are ready to set up the system of two equations.

Step 4: By putting two simple equations from Step 3 one above the other, we obtain a system of linear equations with two variables:

$$\begin{cases} h + w = 240 \\ h - w = 220 \end{cases}$$

Solve this system by the addition method:

$$\begin{cases} h + w = 240 \\ h - w = 220 \end{cases}$$
$$2h \quad\;\; = 460$$

Divide both sides by 2:

$h = 230$ mph

We found that the helicopter's speed in still air is 230 mph.

Step 5: We found the helicopter's speed in still air. To find the wind speed, let's plug this speed into one of the simple equations. We can plug the helicopter's speed (230 mph) into the simple equation for the tailwind travel:

$h + w = 240$

$230 + w = 240$

Isolate the variable and collect like terms on the right:

$w = 10$ mph

We found that the wind's speed is 10 mph.

Solution: The helicopter's speed in still air is 230 mph. The wind's speed is 10 mph.

Practice Problems

Problem 1: Two cars leave two towns at the same time and start traveling toward each other. One averages 50 mph and the other 55 mph. How far does each travel before meeting if the towns are 420 miles apart?

Problem 2: Two hikers are 16.5 miles apart and are walking toward each other. They meet in 3 hours. Find the rate of each hiker if one hiker walks 1.5 mph faster than the other.

Problem 3: Two cars start from the same place and drive in opposite directions. The first car averages 57 mph, while the second car averages 63 mph. In how many hours will the two cars be 600 miles apart?

Problem 4: A truck starts at a speed of 50 mph. A car starts 3 hours later at a speed of 65 mph to overtake the truck. In how many hours will the car overtake the truck?

Problem 5: A boy starts running at a speed of 6 mph. A girl starts on her bike 1 hour later, at a speed of 12 mph. How far will she travel before catching up with the boy?

Problem 6: A bus traveling at 60 mph overtakes a car traveling at 40 mph that had a 2-hour head start. How far from the starting point are the vehicles?

Problem 7: A plane flying with a tailwind flew 300 miles in 2 hours. It took the plane 2.5 hours to fly the same distance against the wind. Find the rate of the wind.

The Least You Need to Know

- In both moving toward and moving in opposite direction problems, the entire distance traveled by the two parties is the sum of their distances.
- In catching-up problems, the head start time should be added to the time of the faster object to equalize the times.
- Traveling with the wind or current increases the rate by the speed of the wind or current, while traveling against decreases it.
- If we have two unknown variables in a problem, we need to set up a system of two equations to find these variables.

The Misfits

In This Chapter

- Applying general strategies and techniques
- Finding similarities with a problem on a familiar topic
- Selecting the major and secondary topic of a word problem
- Writing equations for more challenging problems

We've arrived at the last stop of our journey. In this final chapter, we consider problems that we haven't encountered yet. You might be surprised to learn this and think, "Why did she wait until the last chapter to introduce something new?" Usually, the last chapter in any math book is a summarization and reflection of the previous topics. This is exactly what we will be doing here. You're not going to learn any new techniques or formulas. You're going to learn how to apply our proven methods and strategies to problems that have a little "twist" in them. You'll see what I mean as soon as you start reading the next section.

Problems That Defy Definition

You may encounter a word problem that you can't identify as a specific type of problem we've discussed in this book. This shouldn't discourage or intimidate you, though, since your arsenal of skills for solving algebra word problems has expanded and you're capable of using general strategies, procedures, and reasoning to succeed.

As always, read the problem carefully, making certain that you fully understand the problem and are able to retell it in your own words. Next, choose a meaningful variable to represent an unknown quantity and express other quantities by writing algebraic expressions. Sketch any chart, table, or drawing that might be helpful to visualize the problem. Write the problem's equation and solve it. Be especially alert to answering the correct problem's question.

Another tip is to try to connect a problem at hand with any familiar type of problem we have already discussed.

The first problem tests our ability to choose the right variable and use common sense.

Sample Problem 1

In a barn were horses and people. If we counted 60 heads and 204 legs in the barn, how many horses and how many people were in the barn?

Step 1: Let x represent the number of people. Since both people and horses have just 1 head and together they have 60 heads, the number of horses' heads (and horses) is $60 - x$.

Step 2: Each human has 2 legs, and there are x of them in the barn. The number of human legs can be expressed as $2x$. Each horse has 4 legs, and the number of horses is $60 - x$, so the number of horses' legs can be expressed as $4(60 - x)$.

A WORD OF ADVICE

In many algebra word problems, using common knowledge, like the number of limbs that different species have in Problem 1, helps to solve the problem.

Step 3: We know that the total number of legs in the barn is 204; $2x$ of these legs belong to humans and $4(60 - x)$ of these legs belong to horses. We can write this mathematically as:

$2x + 4(60 - x) = 204$

Distribute on the left side:

$2x + 240 - 4x = 204$

Isolate the variable and collect like terms on both sides:

$2x - 4x = -240 + 204$

$-2x = -36$

Divide both sides by -2:

$x = 18$

We found that there are 18 humans in the barn.

Step 4: To find the number of horses, substitute 18 into the expression for the number of horses from Step 1:

$60 - x = 60 - 18 = 42$

We found that there are 42 horses in the barn.

Solution: There are 18 people and 42 horses in the barn.

The next problem most closely resembles the dry mixture problems that we dealt with in Chapter 11. But instead of the total amount of money, in the next problem we have the total number of acres (the area); and instead of the price per pound, we have the amount of corn per acre. But the reasoning behind the problem is the same: find the total crop from the first field and the total crop from the second field, and then write the equation using the facts from the problem.

Sample Problem 2

The total area of two farm fields is 120 acres. The first field averaged 178 bushels per acre, and the second field averaged 190 bushels per acre. Find the area of each field if the total amount of corn from the first field is 2,960 bushels more than the total amount of corn from the second field.

Step 1: Let's assign variables and represent the area of the first field as x. Then the area of the second field can be expressed as $120 - x$, since the total area is 120 acres.

Step 2: We know that the average crop from the first field is 178 bushels per acre. To find the total crop from the first field, we need to multiply the crop from 1 acre (178 bushels) by the number of acres (x):

The total corn from the first field = $178x$

Step 3: We also know that the average crop from the second field is 190 bushels per acre. To find the total crop from the second field, we need to multiply the crop from 1 acre (190 bushels) by the number of acres ($120 - x$):

The total corn from the second field = $190(120 - x)$

Step 4: To write the equation, we use the problem's fact that the total crop from the first field is 2,960 bushels more than the total amount of corn from the second field:

Total corn from the first field – Total corn from the second field = 2,960

Substitute the two expressions for the total crops from Steps 2 and 3:

$178x - 190(120 - x) = 2,960$

Distribute on the left side:

$178x - 22,800 + 190x = 2,960$

Isolate the variable and collect like terms on both sides:

$178x + 190x = 22,800 + 2,960$

$368x = 25,760$

Divide both sides by 368:

$x = 70$

We found that the area of the first field is 70 acres.

WORTHY TO KNOW

Many algebra word problems aren't based on real-life applications. They are meant to utilize algebra rules, formulas, and strategies and to teach you how to translate plain words into algebraic equations.

Step 5: To find the area of the second field, substitute 70 into the expression from Step 1 for the area of the second field:

$120 - x = 120 - 70 = 50$

We found that the area of the second field is 50 acres.

Solution: The area of the fields is 70 acres and 50 acres.

Problems That Belong to More Than One Type

In this section, we challenge ourselves with problems that can be identified as two types in one. To some extent, it's like having a small side plot that is interwoven into the major plot in a novel. In a certain problem, the main type should be identified first and a general plan for the solution should be worked out. Then, using approaches and strategies for the secondary theme, attend to the secondary topic at hand and express all needed values. Finally, write the final problem's equation as required by the main type.

The first problem in this section is a number type problem mixed with a percent type. We apply strategies we've used in Chapters 7 and 8.

Sample Problem 3

The sum of two numbers is 2,490. Find the two numbers if 6.5% of one number equals 8.5% of another.

Step 1: Let x represent the first number. Then the second number can be expressed as $2,490 - x$, since their sum is 2,490.

Step 2: The next piece of information, "6.5% of one number," means that we need to find a part (6.5%) of a whole using the main percent formula:

$a = p\% \cdot b$

Change percent to decimal form:

$6.5\% = 0.065$

Since we denote the first number as x, we substitute x for the whole into the percent formula:

$a = 0.065 \cdot x$

We found that 6.5% of the first number is $0.065x$.

Step 3: Next, we need to find "8.5% of another number" using the main percent formula as well:

$a = p\% \cdot b$

Change the percent to decimal form:

$8.5\% = 0.085$

Since we expressed the second number as $2,490 - x$, we substitute this expression for the whole into the percent formula:

$a = 0.085(2,490 - x)$

We found that 8.5% of the second number is $0.085(2,490 - x)$.

Step 4: Since the problem states that 6.5% of the first number is equal to 8.5% of the second number, we can express this fact mathematically as:

$0.065x = 0.085(2,490 - x)$

Distribute on the right side:

$0.065x = 211.65 - 0.085x$

Isolate the variable and collect like terms:

$0.065x + 0.085x = 211.65$

$0.15x = 211.65$

Divide both sides by 0.15:

$x = 1,411$

We found that the first number is 1,411.

Step 5: To find the second number, substitute 1,411 into the expression from Step 1:

$2,490 - x = 2,490 - 1,411 = 1,079$

We found that the second number is 1,079.

Solution: The numbers are 1,411 and 1,079.

The next problem combines elements from a geometry problem and a ratio problem, with geometry being the main theme. Hence, we need to utilize our knowledge of geometric formulas and the ability to write the right ratio.

Sample Problem 4

In a triangle PQR, $\angle P$ is equal to 72°. Find the measure of the other two angles if $\angle Q : \angle R = 4:5$.

Step 1: Since the measure of all three angles in any triangle is 180°, we can find the combined measure of $\angle Q$ and $\angle R$ by subtracting 72° from 180°:

$180° - 72° = 108°$

We found that the measure of $\angle Q$ and $\angle R$ is 108°.

A WORD OF ADVICE

Sometimes a problem doesn't provide a sketch for you—you need to draw it yourself. For Problem 4, try to create a sketch using the variables that we assigned and the known values from the problem.

Step 2: Now let the measure of $\angle Q$ be x. Then the measure of $\angle R$ can be expressed as $108° - x$.

Step 3: We also know that the angles' ratio is 4:5. We can now write the ratio that will help us find $\angle Q$ and $\angle R$:

$$\frac{\angle Q}{\angle R} = \frac{4}{5} = \frac{x}{108-x}$$

Cross-multiply:

$5x = 4(108 - x)$

Distribute on the right side:

$5x = 432 - 4x$

Isolate the variable and collect like terms:

$5x + 4x = 432$

$9x = 432$

Divide both sides by 9:

$x = 48$

We found that the $\angle Q$ is 48°.

Step 4: To find the measure of the $\angle R$, substitute 48° into the expression for the $\angle R$ from Step 2:

$108° - x = 108° - 48° = 60°$

We found that the $\angle R$ is 60°.

Solution: The two angles are 48° and 60°.

If there were no percents, the following problem would have been a classic algebra word problem that requires expressing a problem's facts as algebraic expressions and writing the final equation. But the problem includes percents, and this offers an additional twist that requires our attention.

Sample Problem 5

A crew of workers mowed half of a farm plus 2 more acres on the first day. During the second day, they mowed 25% of the remaining area and the last 6 acres. Find the area of the farm.

Step 1: Let x represent the area of the farm. Then half of the area can be represented as $\frac{x}{2}$. The area that the crew mowed during the first day can be expressed as $\frac{x}{2}+2$, since the crew mowed half of the farm plus 2 more acres.

DEAD ENDS

Sometimes a problem tries to trick you when you're looking for a part of the whole: the "whole" you should use is not the real "whole" at all. For instance, in the problem about a crew mowing a farm, we deal with 75% of the whole, which is, in reality, only a part of the farm (75% of the remaining area of the farm).

Step 2: We know that the workers finished mowing the remaining area during the second day. Let y represent this remaining area. We also know that they mowed 25% of the remaining area y and the last 6 acres. Let's try to say this differently: the workers first mowed 25% of the remaining area y, so 75% (100% − 25%) of the remaining area was still there to mow. But we already know that this 75% is 6 acres, since this is the number of acres that was left to mow until the workers finish the whole farm. Let's summarize our reasoning by stating that 6 acres constitute 75% of the remaining area y.

Step 3: Since we know a part (6 acres) and the percent (75%), we can find the whole (the remaining area y) using the main percent formula for the whole:

$$b = \frac{a}{p\%}$$

Convert percents into decimals and substitute the known values into the percent formula:

75% = 0.75

$$y = \frac{6}{0.75} = 8$$

We found that the remaining area that the crew mowed during the second day is 8 acres.

Step 4: Finally, we can write the problem's equation by translating the following statement into a mathematical expression: the area that the crew mowed the first day $\left(\frac{x}{2}+2\right)$ plus the area that the crew mowed during the second day (8 acres) equals the total area of the farm (x):

$$\left(\frac{x}{2}+2\right)+8 = x$$

Step 5: To solve the equation, multiply each term by the least common denominator (LCD), which is 2, to eliminate fractions:

$$\frac{2x}{2} + 2 \cdot 2 + 2 \cdot 8 = 2x$$

Reduce by the common factors and multiply the remaining factors:

$x + 4 + 16 = 2x$

$x + 20 = 2x$

Isolate the variable and collect like terms:

$x - 2x = -20$

$-x = -20$

Divide both sides by −1:

$x = 20$

We found the area of the farm.

Solution: The area of the farm is 20 acres.

The last problem in this chapter is a round-trip motion type problem that also includes percents. We need to utilize strategies that we learned in Chapters 7 and 18.

Sample Problem 6

A car travels from one town to another at a speed of 60 mph. On its way back, the car drives at the same speed for 75% of the distance, and drives at the speed of 40 mph for the rest of the trip. Find the distance between the towns if it took the car 10 more minutes to travel back home.

Step 1: We know that, in the round-trip problem, the distance traveled to the destination is equal to the distance traveled back home. Let *d* represent this distance. To write the problem's equation, we use the fact that the journey back home took 10 minutes longer, since the speed changed.

DEAD ENDS

Be alert that even if we chose the distance as our variable, the equation is about the time of travel back and forth. This is often the case in round-trip motion problems, as we discussed in Chapter 18.

Step 2: Let's first discuss the time it took to travel to the destination. Since the distance is d and the speed is 60 mph, we can use the simple formula for time calculation that we discussed in Chapter 18:

$$t = \frac{d}{r}$$

Substitute the known values and variables:

$$t = \frac{d}{60}$$

We expressed the time of travel to the destination.

Step 3: Let's now deal with the time of travel back home. From the problem, we know that the car drove with the same speed for 75% of the distance; we can express this distance as a part of the whole (the total distance d) using the percent formula and by changing 75% into decimal form:

75% = 0.75

Part of the distance = $p\%$ · (whole) = 0.75d

The time t_1 that the car took to drive this distance at the speed of 60 mph can be expressed as:

$$t_1 = \frac{0.75d}{60}$$

Step 4: Since the total distance is equal to 100% as the whole, the remaining distance can be found by subtracting 75% from 100%:

100% − 75% = 25%

We found that the remaining distance constitutes 25%.

Step 5: We can express this distance as a part (25%) of the whole (the total distance d) using the main percent formula and by changing 25% into decimal form:

25% = 0.25

Part of the distance = $p\%$ · (whole) = 0.25d

The car drove this part of the way at the speed of 40 mph. The time t_2 that the car took to drive this distance at the speed of 40 mph can be expressed as:

$$t_2 = \frac{0.25d}{40}$$

Step 6: The travel back home lasted 10 more minutes. Let's convert 10 minutes into hours using the procedure we discussed in Chapter 18:

$$10 \text{ minutes} = \frac{10}{60} = \frac{1}{6} \text{ hours}$$

We found that 10 minutes is $\frac{1}{6}$ hours.

Step 7: The total time of travel back home is equal to the time driven at the speed of 60 mph plus the time driven at the speed of 40 mph. To equalize the times of travel to the destination and back home, we need to subtract 10 minutes from the total time of travel back home, since the travel to the destination was 10 minutes shorter:

$$t = t_1 + t_2 - \frac{1}{6}$$

Step 8: Substitute the values for the times from Steps 2, 3, and 5 into the preceding equation in Step 7:

$$\frac{d}{60} = \frac{0.75d}{60} + \frac{0.25d}{40} - \frac{1}{6}$$

Multiply each term of the equation by the LCD, which is 120, to eliminate fractions:

$$\frac{120d}{60} = \frac{120(0.75d)}{60} = \frac{120(0.25d)}{40} - \frac{120}{6}$$

Reduce by common factors:

$$\frac{^2\cancel{120}d}{_1\cancel{60}} = \frac{^2\cancel{120}(0.75d)}{_1\cancel{60}} = \frac{^3\cancel{120}(0.25d)}{_1\cancel{40}} - \frac{^{20}\cancel{120}}{_1\cancel{6}}$$

Multiply the remaining factors:

$$2d = 2(0.75d) + 3(0.25d) - 20$$

Multiply on the right side:

$$2d = 1.5d + 0.75d - 20$$

Isolate the variable and collect like terms:

$$2d - 1.5d - 0.75d = -20$$

$$-0.25d = -20$$

Divide both sides by −0.25:

$$d = 80$$

Solution: The distance between the two towns is 80 miles.

Practice Problems

Problem 1: Mary is on a 1,850 calories-per-day diet plan. This plan permits 950 calories less than twice the number of calories permitted by her friend Lorry's diet plan. How many calories are permitted by Lorry's plan?

Problem 2: A car repair bill was $1,074. This included $450 for parts and $52 for each hour of labor. Find the number of hours of labor.

Problem 3: A car dealership sold 62 cars during October 2009. This represented 4 more than twice the number of cars sold during December 2008. How many cars did they sell during December 2008?

Problem 4: In a triangle PQR, $\angle P$ is equal to 37°. Find the measure of the other two angles if $\angle Q{:}\angle R$ = 4:7.

Problem 5: During the first trip, a car used 25% of its gas tank; during the second trip, it used 20% of the rest of the gas tank. After the two trips, the tank still had 4 gallons more than the car used for both trips combined. How many gallons of gas were there at the start?

Problem 6: A triangular lot is enclosed with 135 yards of fencing. The longest side of the lot is 5 yards more than twice the length of the shortest side. The length of the shortest side is 75% of the length of the third side. Find the length of the longest side.

Problem 7: A bus starts traveling to another town that is 160 km away. A car starts out 2 hours later, and the ratio of the speed of the car to the speed of the bus is 3:1. If the car arrives at another town 40 minutes earlier than the bus, find the speed of the car.

The Least You Need to Know

- Solving a problem step by step ensures success.
- When a problem relates to two types, identifying the main type helps to work out the solution plan.
- The strategies and techniques you learned in this book can help solve a variety of word problems.
- All word problems can be conquered by choosing the right variables and using good reasoning.

Glossary

absolute value   The distance of a number from zero on the number line, always positive. For example, $|-5| = 5$, and $|5| = 5$.

addition method   The method of solving a system of linear equations by adding two equations to eliminate one of the variables.

adjacent   Next to; often refers to angles.

algebra   A branch of mathematics that uses variables to express general rules about numbers, operations, and relationships.

algebraic expression   A mathematical statement that contains numbers, variables, and operation symbols. It does not contain an equals sign.

alloy   A homogeneous mixture of two or more elements, at least one of which is a metal, with the resulting material having metallic properties.

angle   Two rays that share an endpoint.

area   The measure, in square units, of the interior region of a two-dimensional object.

arithmetic average   *See* mean.

Associative Property of Addition   The sum stays the same when the grouping of addends is changed: $(a + b) + c = a + (b + c)$.

Associative Property of Multiplication   The product stays the same when the grouping of factors is changed: $a(b \cdot c) = (a \cdot b)c$.

average   *See* mean.

binomial   A polynomial that has only two terms.

Celsius (C)   The metric-system scale to measure temperatures.

coefficient   A number in front of a monomial; for example, the coefficient of $15xy$ is 15.

collecting like terms   Combining terms that have the same variable parts.

common factor   A number that is a factor of two or more numbers.

Commutative Property of Addition   The sum stays the same regardless of the order of addends: $a + b = b + a$.

Commutative Property of Multiplication   The product stays the same regardless of the order of factors: $a \cdot b = b \cdot a$.

complementary angles   Two angles whose combined measure is 90°.

complex fraction   A fraction that has fractions in its numerator, denominator, or both; also called a compound fraction.

compound interest   Interest paid on the original principal and on interest that becomes part of the account.

congruent   Having exactly the same size or shape.

constant   A number that does not have any variables attached to it.

cross-multiplication   A method of finding a missing numerator or denominator in ratios or equivalent fractions by equalizing the cross-products.

decimal   A numeral containing a decimal point which separates the ones from the tenths place. Every numeral that has only a decimal part always has a value less than 1.

degree (angle measure)   A unit for measuring angles.

degree Celsius (°C)   A unit to measure temperature in the Celsius scale.

degree Fahrenheit (°F)   A unit to measure temperature in the Fahrenheit scale.

degree Kelvin (°K)   A unit to measure temperature in the Kelvin scale.

denominator   The number or algebraic expression below the fraction bar in a fraction.

digit   One of the 10 symbols from 0 to 9, inclusive.

distributive property   Multiplying each addend before adding does not change the product: $a(b + c) = ab + ac$.

equation   A statement that two mathematical expressions are equal.

even   A whole number divisible by 2.

expanded form   A way to write a number that shows the place value of each digit; for example, 345 = 3(100) + 4(10) + 5(1).

exponent   The number that tells how many equal factors there are; for example, $2^5 = 2 \cdot 2 \cdot 2 \cdot 2 \cdot 2$.

expression   One or more terms connected by signs of operations.

factor   An integer that can be evenly divided into another.

fraction   A ratio of two integers; a number that has a numerator and a denominator.

greatest common factor (GCF)   The largest number that divides evenly into two or more numbers.

grouping symbols   Symbols such as parentheses or brackets that separate the terms of an algebraic expression into several groups.

improper fraction   A fraction whose numerator is larger than the denominator; for example, $\frac{7}{5}$.

integer   Whole numbers and their opposites: –3, –2, –1, 0, 1, 2, 3 ….

irrational number   Numbers that cannot be written as a ratio of two integers.

least common denominator (LCD)   The smallest common multiple of the denominators of two or more fractions.

leg   One of the two sides that form the right angle in a right triangle.

like terms   Terms that have the same variables with the same exponents.

linear equation   An equation with one variable whose exponent is 1.

mean   The sum of a set of numbers divided by the number of numbers in the set.

negative numbers   Numbers less than zero. They are represented by points to the left of the origin on the real number line.

number line   A diagram that represents numbers as points on a line.

numerator   The top number of a fraction.

odd number   A number not divisible by 2.

opposites   Two points on the real number line that are at the same distance from the origin but are on opposite sides of the origin. For example, 5 and –5 are opposites.

order of operations   Rules describing what sequence to use in evaluating expressions.

parallelogram   A quadrilateral with two pairs of parallel and congruent sides.

percent   A way of expressing hundredths, or a number divided by 100, usually denoted by the symbol %.

place value   The value of the position of a digit in a number. For example, in 742, the 4 is in the tens place, so it stands for 40.

polygon   A closed two-dimensional figure formed by consecutive line segments that meet only at their endpoints.

polynomial   The sum or the difference of distinct terms.

positive   A number that is greater than zero.

power   *See* exponent.

prime factorization   The expression of a number as the product of its prime factors. For example, the prime factorization of 18 is $2 \cdot 3 \cdot 3$.

prime number   A number that has exactly two positive factors, itself and 1. The first prime number is 2.

principal   The amount of money initially deposited into a bank account with simple or compound interest.

product   The result obtained when numbers or expressions are multiplied.

proportion   An equation that equates two ratios.

quadrilateral   A four-sided polygon.

quotient   The result obtained when numbers or expressions are divided.

rational number   A number that can be expressed as a fraction, a terminating decimal, and a repeating decimal.

real numbers   A set of all rational and irrational numbers.

reciprocals   Two numbers that have a product of 1. For example, $\frac{3}{5}$ and $\frac{5}{3}$ are reciprocals because $\frac{3}{5} \cdot \frac{5}{3} = 1$.

rectangle   A quadrilateral with two pairs of congruent, parallel sides and four right angles.

reduce   To put a fraction into simplest form. For example, $\frac{3}{6} = \frac{1}{2}$.

repeating decimal   A decimal that has an infinitely repeating sequence of digits. For example, $6.2323... = 6.\overline{23}$.

set   A well-defined collection of objects.

side   A line segment connected to other segments to form a polygon.

sign   *See* signed number.

signed number   A positive or negative number.

simple interest   Involves interest calculated only on the principal.

solution   Any value(s) for variables that makes an equation(s) true.

solving a proportion   Finding the value for the variable in a proportion.

solving a system of linear equations   Finding the solution of a system of linear equations.

solving an equation   Finding the solutions of an equation.

square   A parallelogram with four congruent sides and four right angles.

substitution   Replacing a variable with an equivalent value or expression.

substitution method   A method of solving a system of two equations when one variable is written in terms of another variable in one equation and the resulting expression is substituted into the other equation.

supplementary angles   Two angles whose combined measure is 180°.

system of equations   Two equations in two variables considered together.

term   A number, variable, product, or quotient in an expression. A term is not a sum or difference. For example, in $2x - 3x + 5$, there are three terms.

terminating decimal   A decimal with a finite number of digits.

unit   A precisely fixed quantity used to measure.

variable   A quantity that can have different values.

vertex   Two rays sharing a common endpoint form an angle. This common endpoint is called the vertex of the angle.

whole number   Any of the numbers 0, 1, 2, 3, and so on.

Equals sign:

Keywords that indicate the equals sign: *equal to, is, was, will be, is, the same as.*

Addition:

Keywords for addition: *sum, total, more than, greater than, plus, older than, increased by, exceeds by.*

The sum of 8 and a number	$8 + x$
Nine more than a number	$y + 9$
A number plus 11	$x + 11$
A number increased by 13	$x + 13$
Olga is 7 years older than Michael	$x + 7$
A number is 6 larger	$x + 6$
Side y is 2 feet longer than side x	$y = x + 2$
A number exceeds another by 2	$x = y + 2$

Subtraction:

Keywords that indicate subtraction: *difference, less than, minus, subtracted from, younger than, shorter than, decreased by.*

A number minus 4	$x - 4$
A number decreased by 6	$x - 6$
A number subtracted from 20	$20 - x$
The difference of 10 and a number	$10 - x$

The difference between a number and 10	$x - 10$
Six less than a number	$x - 6$
Eight minus a number	$8 - y$
19 fewer than a number	$x - 19$
Helen is 8 years younger than Rita	$x - 8$
One side is 3 feet shorter than the other side	$y = x - 3$

Multiplication:

Keywords that indicate multiplication: *product, multiplied by, twice, times, doubled, tripled, of.*

Six times a number	$6x$
The product of 5 and a number	$5x$
Twice a number	$2x$
Fifteen multiplied by a number	$15x$
A number multiplied by another number	xy
Half a number	$\dfrac{1}{2}x$
Three fourths of a number	$\dfrac{3}{4}x$
The quantity is doubled	$2x$
The quantity is tripled	$3y$

Division:

Keywords that indicate division: *quotient, divided by, per, in.*

The quotient of a number and 5	$\dfrac{x}{5}$
Seven divided by a number	$\dfrac{7}{x}$
20 miles in 5 hours	$\dfrac{20}{5}$

More than one operation:

Three more than 4 times a number	$4x + 3$
Three less than 3 times b	$3b - 3$
The difference between three times a number and 13	$3x - 13$
Three times the difference of a number and 10	$3(x - 10)$
Three increased by 12 times a number	$3 + 12x$
The sum of 3 times x and 2 times y	$3x + 2y$
Five less than the product of x and y	$xy - 5$
Six times the sum of x and 2 times y	$6(x + 2y)$
Four more than 5 times a number	$5x + 4$
Three less than the sum of a number and 2	$(x + 2) - 3$
Four times x plus 2 times y	$4x + 2y$
Two times x minus 5 times y	$2x - 5y$
The sum of x and 3 times y	$x + 3y$
Seven times the sum of x and y	$7(x + y)$
Eight less than 3 times the difference of x and 4 times y	$3(x - 4y) - 8$
Fifteen more than 2 times the sum of 5 and a number	$2(5 + x) + 15$
Four times z minus 3 times the sum of x and y	$4z - 3(x + y)$
One third of the sum of 2 times a number plus 5	$\frac{1}{3}(2x + 5)$
Two thirds of the difference of 5 times a number and 2 times another number	$\frac{2}{3}(5x - 2y)$

Rules and Formulas Review

Order of Operations:

Parentheses

Exponents

Multiplication or division

Addition or subtraction

Left-to-right

Relationship Symbols:

= Equal

≠ Not equal

≈ Approximately

> Greater than

< Less than

Absolute Value:

The absolute value of a number is always positive:

$|a| = a$, if $a > 0$

$|a| = -a$, if $a < 0$

FOIL Method:

The product of $(a + b) \cdot (c + d)$ can be written as:

$a \cdot c + a \cdot d + b \cdot c + b \cdot d$

First   Outer   Inner   Last

Percent Formula:

$b = p\% \cdot a$

b is a part of the whole.

$p\%$ is the percent entered in decimal form.

a is the whole.

Work Formula:

$$\frac{1}{t_1} + \frac{1}{t_2} = \frac{1}{t_3}$$

t_1 is the time taken by the first person.

t_2 is the time taken by the second person.

t_3 is the time taken by both working together.

Simple Interest:

$I = Prt$

P: Principal

r: Annual interest rate

t: Time in years

Compound Interest:

$$A = P\left(1 + \frac{r}{n}\right)^{nt}$$

P: Principal

r: Annual rate

t: Time in years

n: Number of compounding per year

Discount:

Original price x

$x - p\% \cdot x =$ Sale price

$p\%$ is the percent of discount in decimal form.

Sale price = (Original price) − (Original price) · (p%)

p% is the percent of discount in decimal form.

$$\frac{\text{discount amount}}{\text{original price}} = \text{discount percent}$$

Lever Law:

$$w_1 \cdot d_1 = w_2 \cdot d_2$$

w_1 and w_2 are weights.

d_1 and d_2 are arms of the weights or distances.

Temperature:

$$°F = \frac{9}{5}C + 32$$

$$°C = \frac{5}{9}\left(F - 32\right)$$

$$°K = °C + 273$$

$$1°F = \frac{5}{9}°C$$

$$1°C = \frac{9}{5}°F$$

°F: Fahrenheit degrees

°C: Celsius degrees

°K: Kelvin degrees

Average (Arithmetic Mean):

$$\text{average} = \frac{\text{sum of terms}}{\text{number of terms}}$$

Sometimes it is convenient to use the same formula in a different form:

Sum of terms = Average · Number of terms

Weighted Average:

$$\text{weighted average} = \frac{\text{sum of weighted terms}}{\text{total number of terms}}$$

Distance–Rate–Time:

Distance: $d = rt$

Rate/speed: $r = \dfrac{d}{t}$

Time: $t = \dfrac{d}{r}$

(d = distance, r = rate/speed, t = time)

Geometry Formulas:

Rectangle:

Area: $A = w \cdot l$

Perimeter: $P = 2l + 2w$

A: Area

P: Perimeter

l: Length

w: Width

Square:

Area: $A = a^2$

Perimeter: $P = 4a$

A: Area

P: Perimeter

a: Length of the side

Triangle:

$A = \dfrac{1}{2}bh$

$P = a + b + c$

A: Area

P: Perimeter

b: Base

h: Height

a, c: Sides

The sum of all three angles equals 180°.

Complementary and Supplementary Angles:

Complementary angles are two angles whose measures have the sum 90°.

Supplementary angles are two angles whose measures have the sum 180°.

Answers to Practice Problems

Chapter 5: Getting Rational with Ratio Problems

Problem 1: Convert the measure of the width into inches: 3 feet, 3 inches = 3(12) + 3 = 39 inches. Let the length be x. Then $\dfrac{\text{width}}{\text{length}} = \dfrac{39}{x} = \dfrac{3}{4}$, 4(39) = 3$x$, 156 = 3$x$, x = 52.

Convert inches into feet and inches: 52:12 = 4 feet plus remainder 4. The length is 4 feet and 4 inches.

Answer: The length is 4 feet and 4 inches.

Problem 2: A man's ratio of red blood cells to all blood cells is: $\dfrac{300}{240{,}000} = \dfrac{1}{800}$. The normal ratio is $\dfrac{1}{5{,}000}$.

Answer: His count is high.

Problem 3: Let the number of dentists who recommended the toothpaste be x. Then $\dfrac{5}{8} = \dfrac{x}{528}$, 8$x$ = 5(528), 8x = 2,640, x = 330.

Answer: 330 dentists recommended the new toothpaste.

Problem 4: Anna's number of green marbles is x and number of red marbles is 40 − x. Obtain the ratio from Lora: $\dfrac{\text{red}}{\text{green}} = \dfrac{45}{30} = \dfrac{3}{2}$. The ratio for Anna: $\dfrac{\text{red}}{\text{green}} = \dfrac{3}{2} = \dfrac{40 - x}{x}$, 2(40 − x) = 3x, 80 − 2x = 3x, 5x = 80, x = 16 green marbles. Lora has 30 − 16 = 14 more green marbles.

Answer: Lora has 14 more green marbles.

Problem 5: The number of women is x, the number of men is $x + 840$.

$\dfrac{\text{men}}{\text{women}} = \dfrac{7}{5} = \dfrac{x + 840}{x}$, $5(x + 840) = 7x$, $5x + 4{,}200 = 7x$, $2x = 4{,}200$, $x = 2{,}100$ women.

The number of men is $2{,}100 + 840 = 2{,}940$; the total enrollment is $2{,}100 + 2{,}940 = 5{,}040$.

Answer: The total enrollment is 5,040 students.

Problem 6: If the number of women is x, then the number of men is $500 - x$. To find the original number of women, solve: $\dfrac{7}{3} = \dfrac{500 - x}{x}$, $7x = 3(500 - x)$, $7x = 1{,}500 - 3x$, $10x = 1{,}500$, $x = 150$. The number of men is $500 - 150 = 350$. The number of additional women is y. $\dfrac{2}{1} = \dfrac{350}{150 + y}$, $350 = 2(150 + y)$, $350 = 300 + 2y$, $2y = 50$, $y = 25$.

Answer: 25 additional women should be accepted.

Problem 7: Let the number of pigs be x; then $\dfrac{\text{pigs}}{\text{hens}} = \dfrac{x}{120} = \dfrac{2}{10}$, $10x = 2(120)$, $10x = 240$, $x = 24$. Let the number of goats be y; then $\dfrac{\text{goats}}{\text{hens}} = \dfrac{y}{120} = \dfrac{3}{10}$, $10y = 3(120)$, $10y = 360$, $y = 36$. There are $36 - 24 = 12$ more goats than pigs.

Answer: There are 12 more goats than pigs on the farm.

Chapter 6: Dealing With Proportions

Problem 1: The number of gallons of additional paint is x. $\dfrac{3}{x} = \dfrac{600}{1{,}800}$, $600x = 3(1{,}800)$, $600x = 5{,}400$, $x = 9$.

Answer: A worker needs 9 more gallons of paint.

Problem 2: The distance traveled in 12 hours is x. $\dfrac{250}{5} = \dfrac{x}{12}$, $5x = 12(250)$, $5x = 3{,}000$, $x = 600$ miles.

Answer: The distance traveled in 12 hours is 600 miles.

Problem 3: The height of the sister is x. Then $\dfrac{5.4}{1.8} = \dfrac{3.3}{x}$, $5.4x = 1.8(3.3)$, $5.4x = 5.94$, $x = 1.1$ m.

Answer: The sister is 1.1 meters tall.

Problem 4: The scale is 1 cm = 65 km. The distance on the map is x. $\dfrac{1}{65} = \dfrac{x}{1{,}950}$, $65x = 1{,}950$, $x = 30$ cm.

Answer: The distance between the two cities on the map is 30 cm.

Problem 5: The number of work-hours for 8 workers is x. $\dfrac{3}{8} = \dfrac{x}{48}$, $8x = 3 \cdot 48$, $8x = 144$, $x = 18$ hours.

Answer: It will take 8 trimmers 18 work-hours to finish the job.

Problem 6: The time for 10 people to complete the job is x. $\dfrac{3}{10} = \dfrac{x}{5}$, $10x = 3 \cdot 5$, $10x = 15$, $x = 1.5$ days.

Answer: It will take 1.5 days for 10 people to complete the job.

Problem 7: The number of days for 6 people to mow lawns is x. $\dfrac{4}{6} = \dfrac{x}{6}$, $6x = 4 \cdot 6$, $6x = 24$, $x = 4$ days.

Answer: 6 people will complete the job in 4 days.

Chapter 7: Percentage Solutions

Problem 1: $2\% = 0.02$, $x = 0.02 \cdot 150 = 3$, $x = 3$.

Answer: The number is 3.

Problem 2: $25\% = 0.25$, $x = \dfrac{85}{0.25} = 340$, $x = 340$.

Answer: The number is 340.

Problem 3: $25\% = 0.25$, $x = \dfrac{\$75,500}{0.25} = \$302,000$.

Answer: The total amount of money is $302,000.

Problem 4: $28\% = 0.28$, $x = \dfrac{35}{0.28} = 125$.

Answer: Maria cared for 125 animals.

Problem 5: $x = \dfrac{480}{640} = 0.75$, $0.75 = 75\%$.

Answer: Sara's score is 75%.

Problem 6: $x = \dfrac{35}{175} = 0.2$, $0.2 = 20\%$.

Answer: The percent of European music was 20%.

Problem 7: $\$400,000 - \$390,000 = \$10,000$

$$\dfrac{\$10,000}{\$400,000} = \dfrac{1}{40} = 0.025.$$

$0.025 = 2.5\%$.

Answer: The price drop is 2.5%.

Chapter 8: The Numbers Game

Problem 1: Let the number be x; $8x = 7x + 14$, $x = 14$.

Answer: The number is 14.

Problem 2: The first number is x, the second number is $3x$, the third number is $4x$. $3x + x + 4x = 152$, $8x = 152$, $x = 19$ (the first number), the second is $3x = 3(19) = 57$, the third one is $4x = 4(19) = 76$.

Answer: The numbers are 19, 57, and 76.

Problem 3: The numbers are: x, $(x + 1)$, $(x + 2)$, $(x + 3)$, and $(x + 4)$. $x + (x + 1) + (x + 2) + (x + 3) +(x + 4) = 65$, $5x + 10 = 65$, $5x = 55$, $x = 11$.

Answer: The numbers are 11, 12, 13, 14, and 15.

Problem 4: The four integers are: x, $x + 1$, $x + 2$, $x + 3$. $x + (x + 3) = 49$, $2x + 3 = 49$, $2x = 46$, $x = 23$. Integers are: 23, 24, 25, and 26.

Answer: The third integer is 25.

Problem 5: The odd integer is x, the next even integer is $x + 1$. $3x = 2(x + 1) + 11$, $3x = 2x + 2 + 11$, $x = 13$. The even integer is 14.

Answer: The numbers are 13 and 14.

Problem 6: The three integers are: x, $x + 2$, $x + 4$. $3[x + (x + 2)] = 5(x + 4) + 3$, $3(2x + 2) = 5x + 20 + 3$, $6x + 6 = 5x + 23$, $6x - 5x = 23 - 6$, the first integers is $x = 17$, the second integer is $x + 2 = 17 + 2 = 19$, the third integer is $x + 4 = 17 + 4 = 21$.

Answer: The integers are 17, 19, and 21.

Problem 7: The tens digit is x, the units digit is $x + 3$, and the hundreds digit is $11 - [x + (x + 3)] = 11 - (2x + 3) = 11 - 2x - 3 = 8 - 2x$.

The original number: $100(8 - 2x) + 10x + 1(x + 3)$

The changed number: $100(x + 3) + 10x + 1(8 - 2x)$

The equation: $100(8 - 2x) + 10x + 1(x + 3) = 100(x + 3) + 10x + 1(8 - 2x) - 99$

$-189x + 803 = 108x + 209$, $-297x = -594$, $x = 2$ (this is the tens digit). The units digit is $x + 3 = 2 + 3 = 5$; the hundreds digit is $8 - 2x = 8 - 2(2) = 8 - 4 = 4$.

Answer: The number is 425.

Chapter 9: Age Is Relative

Problem 1: The age today is x, the age in 8 years: $x + 8$. $x + 8 = 2x$, $-x = -8$, $x = 8$.

Answer: Catherine is 8 years old.

Problem 2: The age today is x, the age 4 years ago: $x - 4$, the age in 32 years: $x + 32$.

The equation is: $x - 4 = \dfrac{1}{3}(x + 32)$, $3(x - 4) = x + 32$, $3x - 12 = x + 32$, $2x = 44$, $x = 22$.

Answer: Bill is 22 years old.

Problem 3: The sister's age is x, Ashley's age is $x + 10$. Six years from now, the sister's age: $x + 6$, Ashley's age: $x + 16$.

$x + 16 = 2(x + 6)$, $x + 16 = 2x + 12$, $-x = -4$, $x = 4$. Ashley's age now: $x + 10 = 4 + 10 = 14$.

Answer: Ashley is 14 years old now.

Problem 4: The brother's age: x, Jason's age: $x + 10$. Eight years ago, the brother's age: $x - 8$, Jason's age: $x + 10 - 8 = x + 2$. The equation is: $(x - 8) + (x + 2) = 58$, $2x = 64$, the brother's age: $x = 32$, Jason's age: $32 + 10 = 42$.

Answer: Jason is 42 years old and the brother is 32 years old.

Problem 5: Paul's age: x, Mark's age: $2x + 11$. Five years ago, Paul's age: $x - 5$, Mark's age: $2x + 11 - 5 = 2x + 6$. The equation is: $(x - 5) + (2x + 6) = 52$, $3x = 51$, Paul's age: $x = 17$, Mark's age: $2x + 11 = 2(17) + 11 = 45$.

Answer: Mark is 45 years old.

Problem 6: The daughter's age: x, the mom's age: $2x$. Twelve years ago, the daughter's age: $x - 12$, the mom's age: $2x - 12$.

The equation is: $x - 12 = \dfrac{1}{3}(2x - 12)$, $3(x - 12) = 2x - 12$, $3x - 2x = 36 - 12$; the daughter's age $x = 24$, the mom's age: $2x = 2(24) = 48$.

Answer: The daughter is 24 years old and the mom is 48 years old.

Problem 7: Danny's age now: x, his age 2 years ago: $x - 2$. The uncle's age two years ago: $5(x - 2)$. The uncle's age now: $5(x - 2) + 2 = 5x - 10 + 2 = 5x - 8$. In four years, Danny's age: $x + 4$, the uncle's age: $5x - 8 + 4 = 5x - 4$. The equation is:

$x + 4 = \dfrac{1}{3}(5x - 4)$, $3(x + 4) = 5x - 4$, $3x - 5x = -12 - 4$, $-2x = -16$, Danny's age now: $x = 8$, the uncle's age now: $5x - 8 = 5(8) - 8 = 40 - 8 = 32$.

Answer: The uncle's age now is 32.

Chapter 10: Working It

Problem 1: The time to finish the job together is x.

$\frac{1}{30} + \frac{1}{45} = \frac{1}{x}$. The LCD = 90x. $\frac{90x}{30} + \frac{90x}{45} = \frac{90x}{x}$; $3x + 2x = 90$, $5x = 90$, $x = 18$.

Answer: The two of them will finish the job in 18 minutes.

Problem 2: The time to finish the job by the second worker is x. The equation:

$\frac{1}{16} + \frac{1}{x} = \frac{1}{12}$, the LCD = 48x. $\frac{48x}{16} + \frac{48x}{x} = \frac{48x}{12}$, $3x + 48 = 4x$, $x = 48$.

Answer: The second person alone will finish the job in 48 hours.

Problem 3: The rate is $\frac{1}{12}$. In 10 hours: $10\left(\frac{1}{12}\right) = \frac{10}{12} = \frac{5}{6}$.

Answer: The mechanic can do $\frac{5}{6}$ of the job in 10 hours.

Problem 4: Amy can clean the kitchen in x hours; then Anna can clean the kitchen

in $2x$ hours. The equation: $\frac{1}{x} + \frac{1}{2x} = \frac{1}{40}$, the LCD = 40x, $\frac{40x}{x} + \frac{40x}{2x} = \frac{40x}{40}$, $40 + 20 =$

x, $x = 60$ minutes for Amy or 1 hour; $2x = 120$ minutes, or 2 hours, for Anna.

Answer: It would take 1 hour for Amy to clean, and 2 hours for Anna to clean.

Problem 5: The time to finish the job together is x. Fred's time is $4\frac{1}{2} = \frac{9}{2}$ hours; his

rate is: $\frac{1}{\frac{9}{2}} = \frac{2}{9}$. The brother-in-law's time is 4 hours; his rate is $\frac{1}{4}$. The nephew's time

is $3\frac{1}{2} = \frac{7}{2}$ hours; his rate is $\frac{1}{\frac{7}{2}} = \frac{2}{7}$. The equation is: $\frac{2}{9} + \frac{1}{4} + \frac{2}{7} = \frac{1}{x}$. The LCD is 252x,

$\frac{252x \cdot 2}{9} + \frac{252x \cdot 1}{4} + \frac{252x \cdot 2}{7} = \frac{252x \cdot 1}{x}$. After reducing by common factors: $28x(2) + 63x$

$+ 36x(2) = 252$, $56x + 63x + 72x = 252$, $191x = 252$, $x = \frac{252}{191} = 1\frac{61}{191}$ hours.

Answer: Together they will finish the job in $1\frac{61}{191}$ hours.

Problem 6: The time to empty when working together is x. The rate to fill is $\frac{1}{24}$;

the rate to drain is $\frac{1}{20}$, the rate to empty together is $\frac{1}{x}$. The equation is: $\frac{1}{20} - \frac{1}{24} = \frac{1}{x}$.

The LCD $= 120x$, $\dfrac{120x}{20} - \dfrac{120x}{24} = \dfrac{120x}{x}$; $6x - 5x = 120$, $x = 120$ minutes $= 2$ hours.

Answer: The tank can be emptied in 2 hours.

Problem 7: The time to fill the tank by three pipes is x, the rate is $\dfrac{1}{x}$. The rate to fill by the first pipe is $\dfrac{1}{12}$, the rate to fill by the second pipe is $\dfrac{1}{8}$, the rate to drain by the third pipe is $\dfrac{1}{10}$. The equation is: $\dfrac{1}{12} + \dfrac{1}{8} - \dfrac{1}{10} = \dfrac{1}{x}$, the LCD $= 120x$, $\dfrac{120x}{12} + \dfrac{120x}{8} - \dfrac{120x}{10} = \dfrac{120x}{x}$, $10x + 15x - 12x = 120$, $13x = 120$, $x = \dfrac{120}{13} = 9\dfrac{3}{13}$ minutes.

Answer: The time to fill the tank when three pipes are working together is $9\dfrac{3}{13}$ minutes.

Chapter 11: Coins and Mixtures

Problem 1: The number of dimes is d, the number of quarters is $3d + 6$. The equation: $10d + 25(3d + 6) = 830$, $10d + 75d + 150 = 830$, $85d = 680$, the number of dimes: $d = 8$, quarters: $3d + 6 = 30$.

Answer: Tom has 8 dimes and 30 quarters.

Problem 2: The number of dimes is d, the number of nickels is $3d$, and the number of quarters is $d + 7$. The equation is: $5(3d) + 10d + 25(d + 7) = 425$, $50d = 250$, dimes $d = 5$, nickels $= 3(5) = 15$, and quarters $= 5 + 7 = 12$.

Answer: There are 5 dimes, 15 nickels, and 12 quarters.

Problem 3: The number of 34¢ stamps is x, the number of 5¢ stamps is $30 - x$. The equation is: $34x + 5(30 - x) = 498$, $34x + 150 - 5x = 498$, $29x = 348$, 34¢ stamps: $x = 12$, 5¢ stamps: $30 - 12 = 18$.

Answer: She bought 12 stamps for 34¢ and 18 stamps for 5¢.

Problem 4: The number of nickels is n and the number of dimes is d. $n + d = 26$; revised: $n = 26 - d$. The equation: $5n + 10d = 180$. Substitute the revised expression: $5(26 - d) + 10d = 180$, $130 - 5d + 10d = 180$, $5d = 50$, $d = 10$ dimes, $n = 26 - 10 = 16$ nickels.

Answer: He has 16 nickels and 10 dimes.

Problem 5: The amount of the expensive tea is x, the amount of the less expensive tea is $100 - x$. The equation is: $6x + 4(100 - x) = 5(100)$, $6x + 400 - 4x = 500$, $2x = 100$, $x = 50$ pounds of the expensive tea, the less expensive tea: $100 - 50 = 50$ pounds.

Answer: 50 pounds of the expensive tea and 50 pounds of the less expensive tea are required.

Problem 6: The amount of white chocolate candies is x, the amount of mixture is $x + 20$. The equation is: $(448)x + (896)(20) = 728(x + 20)$, $448x + 17,920 = 728x + 14,560$, $-280x = -3,360$, $x = 12$ pounds.

Answer: 12 pounds of white chocolate candies must be mixed.

Problem 7: The number of tickets for children is x, the number of adult tickets: $280 - x$. The equation is: $8x + 18(280 - x) = 4,590$, $8x + 5,040 - 18x = 4,590$, $-10x = -450$, $x = 45$ child tickets, the number of adult tickets is $280 - 45 = 235$.

Answer: 45 child tickets and 235 adult tickets were sold.

Chapter 12: Manage Your Finances

Problem 1: Convert months into years: 42 months = 36 months + 6 months = 3.5 years. Change the rate into decimal form: 6% = 0.06. Use the simple interest formula and substitute $300 for the interest: $I = Prt$, $300 = P(0.06)(3.5) = P(0.21)$. Divide both sides by (0.21): $P = \dfrac{300}{0.21} \approx 1,428.57$ (round to the hundredth place).

Answer: The principal is $1,428.57.

Problem 2: The principal is $800, the rate in decimal form is 5% = 0.05, and $t = 4$. Use the simple interest formula to find the interest amount: $I = Prt = (800)(0.05)(4) = \$160$. To find the total amount to be paid off, add the principal and the earned interest: Total = $800 + $160 = $960.

Answer: The total amount of money to be paid off is $960.

Problem 3: Let x be the rate on the first account; then the rate on the second account is $x - 0.0075$, since the second rate is 0.75% less and 0.75% = 0.0075 in decimal form. Time $t = 1$. Interest earned on the first account is $I_1 = 16,000(x)(1) = 16,000x$. Interest earned on the second account is $I_2 = 24,000(x - 0.0075)(1) = 24,000(x - 0.0075)$. Since the first account earned $300 less, the equation is: $24,000(x - 0.0075) - 16,000x = 300$, $24,000x - 180 - 16,000x = 300$, $8,000x = 480$, $x = 0.06$. This is the rate on the first account and in percent form is 0.06 = 6%; therefore, the rate on the second account is $x - 0.75\% = 6\% - 0.75\% = 5.25\%$.

Answer: The rate on the first account is 6% and on the second is 5.25%.

Problem 4: The principal is P = $5,000, the rate is 8.5% – 0.085, t = 3 years, and n = 4. Use the compound interest formula, substitute the values, and calculate the total amount:

$$A = 5,000\left(1+\frac{0.085}{4}\right)^{4\cdot3} = 5,000(1,02125)^{12} \approx 5,000(1,2870) \approx 6,435.09 \text{ (rounded to the}$$

hundredth place).

Answer: The amount of money after three years will be $6,435.09.

Problem 5: Let the principal be x, the total amount is $8,000, the rate is 5% = 0.05, t = 2 years, and n = 12. Use the compound interest formula and substitute the variable and values:

$$8,000 = x\left(1+\frac{0.05}{12}\right)^{12\cdot2} \approx x(1,00417)^{24} \approx x(1.1049). \text{ To find } x, \text{ divide by } 1.1049:$$

$$x \approx \frac{8,000}{1.1049} \approx 7,240.47.$$

Answer: Kelly needs to put in around $7,240.47.

Problem 6: The amount of money invested in the second mortgage account is x, the amount of money invested in the first mortgage account is $2x$. Percents in decimal form are: 3.5% = 0.035 and 8% = 0.08. Interest earned in the first account is I_1 = Prt = $2x(0.035)(1)$ = $0.07x$. Interest earned in the second account is I_2 = $x(0.08)(1)$ = $0.08x$. Total interest of $870 can be expressed as the sum of two interests: $I_1 + I_2$ = $0.07x$ + $0.08x$ = 870, $0.15x$ = 870. Multiply both sides by 100: $15x$ = 87,000, x = $5,800. This is the amount of money to invest into the second account. Money to invest in the first account: $2x$ = $2(5,800)$ = $11,600.

Answer: The person must invest $11,600 and $5,800.

Problem 7: Let the amount of money invested at a 15% return be x, and the amount of money invested at a 12% loss be $20,000 – x$. Percents in decimal form are: 15% = 0.15 and 12% = 0.12. The profit from the first account is $(0.15)x$, and the loss can be expressed as $(0.12)(20,000 – x)$. Profit – Loss = Net gain. $0.15x – 0.12(20,000 – x)$ = 840, $0.15x – 2,400 + 0.12x$ = 840. Multiply both sides by 100: $15x – 240,000 + 12x$ = 84,000, $27x$ = 324,000. x = $12,000 is the money invested at 15%. $20,000 – x$ = $20,000 – 12,000$ = $8,000 is the money invested at a 12% loss.

Answer: One investment was $12,000 and the other was $8,000.

Chapter 13: Deep Discount

Problem 1: The original price is x. $100\% - 25\% = 75\%$, $75\% = 0.75$, $x = \dfrac{\$690}{0.75} = \$920$.

Answer: The original price is $920.

Problem 2: The original price is x. $30\% = 0.30$.

$x - xp\% = \$35$, $x - x(0.30) = \$35$ (multiply by 100).

$100x - 30x = \$3,500$, $70x = \$3,500$, $x = \$50$.

Answer: The original price is $50.

Problem 3: The sale price is x. The decreased percent is $100\% - 16\% = 84\%$; in decimal form: $84\% = 0.84$. Use the formula to calculate the sale price: $x = \$45(0.84) = \$37.80$.

Answer: The sale price is $37.80.

Problem 4: Let the sale price be x. Convert the discount percent into decimal form: $15\% = 0.15$. Use the formula to find the sale price: Sale price = original price – (original price)($p\%$).

$x = 2,520 - 2,520(0.15) = 2,520 - 378 = \$2,142$.

Answer: The sale price is $2,142.

Problem 5: Let the wholesale (original) price be x. Convert 25% into decimal form: $25\% = 0.25$. Since the item is marked up, we need to add the increased amount. Use the formula: $x + p\%x = $ retail price. $x + x(0.25) = \$16.25$. Multiply both sides by 100: $100x + 25x = 1,625$, $125x = 1,625$, $x = \$13$.

Answer: The wholesale price is $13.

Problem 6: Find the saved amount of money: $\$850 - \$637.50 = \$212.50$. To find the discount percent $p\%$, use the formula: $p\% = \dfrac{\text{saved amount}}{\text{original price}} = \dfrac{212.50}{850} = 0.25$. Convert into percents: $0.25 = 25\%$.

Answer: The discount rate is 25%.

Problem 7: Find the decreased percent for the first discount: $100\% - 60\% = 40\%$; in decimal form: $40\% = 0.40$. The sale price $x = $ (original price) · (decreased percent). Substitute the known values: $x = (180) \cdot (0.40) = \$72$. The sale price after the first discount is $72, and this is the new original price.

Find the decreased percent after the second discount: $100\% - 15\% = 85\%$, in decimal form: $85\% = 0.85$, the sale price $y = (72) \cdot (0.85) = \$61.20$.

Answer: The final sale price is $61.20.

Chapter 14: When Algebra Helps Geometry

Problem 1: The first angle is x, the second angle is $8x + 27$. The sum of two complementary angles is $90°$: $x + 8x + 27 = 90$, $9x = 63$, $x = 7°$, the other angle is $90° - 7° = 83°$.

Answer: The two angles are 7° and 83°.

Problem 2: The first angle is x, the second angle is $4x - 5$. The sum of two supplementary angles is $180°$: $x + (4x - 5) = 180$, $5x - 5 = 180$, $5x = 185$, the first angle is $x = 37$, the second angle is $180 - 37 = 143$.

Answer: Angles are 37° and 143°.

Problem 3: The first angle is x, the second angle is $x + 10$, the third angle is $3(x + 10) + 5$. The sum of all three angles is $180°$: $x + (x + 10) + 3(x + 10) + 5 = 180$, $5x + 45 = 180$, $5x = 135$, the first angle is $x = 27$, the second angle is $x + 10 = 27 + 10 = 37$, the third angle is $3(x + 10) + 5 = 3(27 + 10) + 5 = 3(37) + 5 = 116$.

Answer: Angles are 27°, 37°, and 116°.

Problem 4: The first side is x, the second side is $x + 2$, the third side is $x - 10$. The perimeter is: $x + (x + 2) + (x - 10) = 70$, $3x - 8 = 70$, $3x = 78$. The first side is $x = 26$, the second is $x + 2 = 26 + 2 = 28$, the third side is $x - 10 = 26 - 10 = 16$.

Answer: The sides are 16, 26, and 28.

Problem 5: The width is x, the length is $3x - 7$. The perimeter is $2w + 2l = 2x + 2(3x - 7) = 66$, $2x + 6x - 14 = 66$, $8x = 66 + 14$, $8x = 80$. The width is $x = 10$, the length is $3(10) - 7 = 30 - 7 = 23$. The area is $A = w \cdot l = 10(23) = 230$.

Answer: The area is 230 square feet.

Problem 6: The side of the square is x, the area is: $A = x^2$. The length of the rectangle is $x + 12$, the width is $x - 8$. The area of the rectangle: $A = (x + 12)(x - 8) = x^2 + 4x - 96$. The areas are the same: $x^2 = x^2 + 4x - 96$, $x^2 - x^2 - 4x = -96$, $-4x = -96$, the side of the square $x = 24$.

Answer: The side of the square is 24 inches.

Problem 7: The width of the building is x, the length of the building is $2x - 15$. The area of the building is $A = x(2x - 15) = 2x^2 - 15x$. The width of the building with the sidewalk is $x + 16$ (plus 8 for each side), the length of the building with the sidewalk is $2x - 15 + 16 = 2x + 1$. The area of the building with the sidewalk is $A_2 = (x + 16)(2x + 1) = 2x^2 + 33x + 16$. The area of the sidewalk is $A_2 - A_1 = 1,216$, $2x^2 + 33x + 16 - (2x^2 - 15x) = 1,216$, $2x^2 + 33x + 16 - 2x^2 + 15x = 1,216$, $48x = 1,216 - 16$, $48x = 1,200$; the width is $x = 25$ feet, the length is $2x - 15 = 2(25) - 15 = 50 - 15 = 35$.

Answer: The width of the building is 25 feet and its length is 35 feet.

Chapter 15: Lever and Temperature Problems

Problem 1: The weight to balance a 400-pound weight is x.

$x(6) = 400(24)$, $6x = 9,600$, $x = 1,600$.

Answer: 1,600 pounds will make a balance.

Problem 2: The distance from the fulcrum to the 160-pound rock is x; the distance from the fulcrum to the gardener is $8 - x$.

$160x = 32(8 - x)$, $160x = 256 - 32x$, $192x = 256$, $x = \dfrac{256}{192} = \dfrac{4}{3} = 1\dfrac{1}{3}$ feet.

Answer: The fulcrum is $1\dfrac{1}{3}$ feet from a 160-pound rock.

Problem 3: The distance from the fulcrum to the fourth boy is x.

The equation is: $85(7) + 60(5) = 55(7) + 102x$, $595 + 300 = 385 + 102x$, $895 - 385 = 102x$, $x = 5$ feet.

Answer: The distance is 5 feet.

Problem 4: $°C = \dfrac{5}{9}(F - 32) = \dfrac{5}{9}(77 - 32) = \dfrac{5}{9}(45) = 25°C.$

Answer: The temperature is 25°C.

Problem 5: $°F = \dfrac{9}{5}C + 32 = \dfrac{9}{5}(85) + 32 = 153 + 32 = 185°F.$

Answer: The temperature is 185°F.

Problem 6: The Celsius reading: $°C = \dfrac{5}{9}(95 - 32) = \dfrac{5}{9}(63) = 35°C.$

Find the Kelvin reading: $°K = °C + 273 = 35°C + 273 = 308°K.$

Answer: The Kelvin reading is 308°K.

Problem 7: We know that $1°F = \dfrac{5}{9}C$; then $63°F = \dfrac{5}{9}(63) = 35°C$.

Answer: The drop on the Celsius scale is 35°C.

Chapter 16: Liquid Solution Problems

Problem 1: The amount of a 35% solution is x, so the amount of a 10% solution is $30 - x$, since the resulting amount is 30 ounces. Percents in decimal form: 35% = 0.35, 10% = 0.10, and 20% = 0.20. The equation is: $0.35x + 0.10(30 - x) = (0.20)(30)$, $0.35x + 3 - 0.10x = 6$. Multiply by 100: $35x + 300 - 10x = 600$, $25x = 300$, $x = 12$ ounces of a 35% solution. The amount of the second solution is $30 - x = 30 - 12 = 18$ ounces.

Answer: 12 ounces of a 35% solution should be mixed with 18 ounces of a 10% solution.

Problem 2: The sodium percentage in the final solution is x. The amount of the resulting solution is $4 + 6 = 10$ liters. Percents in decimal form: 30% = 0.30, 50% = 0.50. The equation is: $4(0.30) + 6(0.50) = 10x$, $1.2 + 3 = 10x$, $4.2 = 10x$, $x = 0.42$. Change decimals into percents: 0.42 = 42%.

Answer: The percentage of sodium in the final mixture is 42%.

Problem 3: The amount of a 20% solution is x, the amount of a 60% solution is $x - 36$. The amount of the resulting solution is $x + (x - 36) = 2x - 36$. Percents in decimal form: 20% = 0.20, 60% = 0.60, and 34% = 0.34. The equation is: $0.20x + 0.60(x - 36) = 0.34(2x - 36)$, $0.20x + 0.60x - 21.6 = 0.68x - 12.24$, $0.20x + 0.60x - 0.68x = 21.6 - 12.24$. Multiply by 100: $20x + 60x - 68x = 2,160 - 1,224$, $12x = 936$, $x = 78$ ounces of a 20% solution is needed. The amount of the second solution is $x - 36 = 78 - 36 = 42$. The total amount of the mixture: $78 + 42 = 120$ ounces.

Answer: There are 120 ounces of the total mixture.

Problem 4: The amount of pure water to be added is x. The amount of the resulting solution is $300 + x$. Percents in decimal form: 30% = 0.30, 20% = 0.20. The equation is about the acid amount: $0.30(300) = 0.20(300 + x)$, $90 = 60 + 0.20x$, $30 = 0.20x$, $x = 150$ liters.

Answer: 150 liters of pure water must be added.

Problem 5: The amount of the first alloy is x, the amount of the second alloy is 24.8 $- x$. Percents in decimal form: 50% = 0.50, 60% = 0.60, and 40% = 0.40. The equation is: $0.60x + 0.40(24.8 - x) = (0.50)(24.8)$, $0.60x + 9.92 - 0.40x = 12.4$. Multiply by 100: $60x + 992 - 40x = 1,240$, $20x = 248$, $x = 12.4$ ounces of the first alloy, the amount of the second alloy is $24.8 - 12.4 = 12.4$ ounces.

Answer: He should use 12.4 ounces of the first alloy and 12.4 ounces of the second alloy.

Problem 6: The amount of the removed water is x, the amount of the resulting solutions will be $90 - x$. The percentage of water in the original solution is 100% − 15% = 85%; the percentage of water in the final solution is 100% − 20% = 80%. Since we are removing pure water, the percentage of water is 100%, or just 1 in decimal form. Percents in decimal form: 85% = 0.85, 80% = 0.80. The equation is: $0.85(90) - 1x = 0.80(90 - x)$, $76.5 - x = 72 - 0.80x$, $-0.20x = -4.5$, $x = 22.5$.

Answer: 22.5 ounces of water must be removed.

Problem 7: The amount of a 20% antifreeze solution removed is x, the amount of pure antifreeze that replaced that amount is also x. The amount of the original and final solution is the same and equals 48 gallons. The percentage of antifreeze in the original solution is 20%, and in the final is 40%. The percentage of the removed antifreeze is also 20% and of added antifreeze is 100% (pure antifreeze), or just 1 in decimal form. Percents in decimal form: 20% = 0.20, 40% = 0.40. The equation is: $0.20(48) - 0.20x + 1x = 0.40(48)$, $9.6 + 0.80x = 19.2$, $960 + 80x = 1,920$, $80x = 960$, $x = 12$ gallons.

Answer: 12 gallons of solution were removed and replaced with pure antifreeze.

Chapter 17: Is Arithmetic Mean?

Problem 1: The score on Sam's fourth test is x. The sum of three scores = average · 3 (the number of scores) = 83(3) = 249; the sum of four scores = average · 4 (the number of scores) = 86(4) = 344. x = sum of four scores − sum of three scores = 344 − 249 = 95.

Answer: He must get at least 95 on his fourth test.

Problem 2: The sum of all terms: $(x - 3) + (2x - 4) + (2x - 2) + (1 - x) + x + (x - 4) = 6x - 12$. Average $= \dfrac{6x - 12}{6} = \dfrac{6(x - 2)}{6} = x - 2$.

Answer: The average is $x - 2$.

Problem 3: The added number is x. The sum of five numbers = average · (number of terms) = 30 · 5 = 150, the sum of six numbers = average · (number of terms) = 32 · 6 = 192. x = sum of six numbers – sum of five numbers = 192 – 150 = 42.

Answer: The added number is 42.

Problem 4: The price for the fifth type of candies is x, the price for the fourth type of candies is $x – 3$. Sum of all terms = average price · number of types = 7 · 5 = 35. Find the sum of all terms: $5 + 7 + 8 + (x – 3) + x = 2x + 17$, then $2x + 17 = 35$, $2x = 18$, $x = \$9$ is the price for the fifth type, the price for the fourth type is $x – 3 = 9 – 3 = \$6$.

Answer: The price per pound for the fourth type is \$6 and for the fifth is \$9.

Problem 5: The number of pounds of all teas sold is 100 + 60 + 40 + 50 = 250.

$$\text{Average} = \frac{8(100) + 9(60) + 7(40) + 8(50)}{250} = \frac{2{,}020}{250} = 8.08.$$

Answer: The average price per pound was \$8.08.

Problem 6: Convert percents into decimal form: 10% = 0.10, 20% = 0.20, 30% = 0.30, 40% = 0.40.

$$\text{Average} = \frac{\text{sum of all terms}}{\text{total weights}} = \frac{0.10(92) + 0.20(68) + 0.30(75) + 0.40(94)}{0.10 + 0.20 + 0.30 + 0.40} =$$

$$= \frac{9.2 + 13.6 + 22.5 + 37.6}{1} = 82.9.$$

Answer: Morris's average grade is 82.9.

Problem 7: $\text{Average} = \dfrac{\text{sum of all terms}}{\text{number of all credits}} = \dfrac{4(4) + 3(3) + 3(2) + 2(4)}{12} = \dfrac{39}{12} = 3.25.$

Answer: Maria's GPA is 3.25.

Chapter 18: Motion Problems with a Single Traveler

Problem 1: The time of travel is 5.5 hours, $d = r \cdot t$, d = 130 mph · 5.5h = 715 miles.

Answer: The distance is 715 miles.

Problem 2: 2h 30 min = $2\dfrac{1}{2} = \dfrac{5}{2}$ hours, $r = \dfrac{d}{t} = d \div t = 15 \div \dfrac{5}{2} = 6\,\text{mph}.$

Answer: The average speed is 6 mph.

Problem 3: The time for the first part of the trip is $t_1 = \dfrac{40\ \text{miles}}{20\ \text{mph}} = 2\ \text{h}$, the distance for the second part of the trip: $60 - 40 = 20$ miles. The time for the second part of the trip is $t_2 = \dfrac{20\ \text{miles}}{10\ \text{mph}} = 2\text{h}$, $t = t_1 + t_2 = 2\ \text{h} + 2\ \text{h} = 4\ \text{h}$. The average speed:

$r = \dfrac{d}{t} = \dfrac{60\ \text{miles}}{4\ \text{h}} = 15\ \text{mph}.$

Answer: The average speed is 15 mph.

Problem 4: The time of the whole trip is 2 hours + 3 hours = 5 hours, $d_1 = 5$ mph · (2h) = 10 miles, $d_2 = 10$ mph · (3h) = 30 miles, $d = d_1 + d_2 = 10 + 30 = 40$ miles, the average speed: $r = \dfrac{d}{t} = \dfrac{40\ \text{miles}}{5\ \text{h}} = 8\ \text{mph}.$

Answer: The average speed is 8 mph.

Problem 5: The distance is d, the total time is $t_1 + t_2 = 6.5$ hours, $\dfrac{d}{30} + \dfrac{d}{35} = 6.5$ (multiply each term by the LCD 210 and reduce by common factors), $7d + 6d = 1{,}365$, $13d = 1{,}365$, $d = 105$ miles.

Answer: George rode out 105 miles.

Problem 6: The actual time of travel is: 8 hours − 2 hours = 6 hours, d is the distance, $\dfrac{d}{2} + \dfrac{d}{4} = 6$ (multiply by the LCD 4 and reduce by common factors), $2d + d = 24$, $3d = 24$, $d = 8$ miles.

Answer: The distance is 8 miles.

Problem 7: The distance is d, the total time is $t_1 + t_2 = 5$ hours, $\dfrac{d}{9} + \dfrac{d}{6} = 5$ (multiply by the LCD 18 and reduce by common factors), $2d + 3d = 90$, $5d = 90$, $d = 18$ miles.

Answer: It is 18 miles from Marilyn's home.

Chapter 19: Motion Problems with Multiple Travelers

Problem 1: The time of travel is t. The distance between two towns is d. $d_1 + d_2 = d$, $50t + 55t = 420$, $105t = 420$, $t = 4$ hours. $d_1 = 50(4) = 200$ miles, $d_2 = 55(4) = 220$ miles.

Answer: The distance that each traveled is 200 miles and 220 miles.

Problem 2: The speed of the first hiker is r, the speed of the second one is $r + 1.5$. The total distance is d. $d_1 + d_2 = d$. The time of travel is 3 hours. $3r + 3(r + 1.5) = 16.5$; $6r = 12$, $r = 2$ mph. The second hiker: $r + 1.5 = 2 + 1.5 = 3.5$ mph.

Answer: The rates are 2 mph and 3.5 mph.

Problem 3: t is the unknown time. The total distance is $d_1 + d_2 = d$. $57t + 63t = 600$, $120t = 600$, $t = 5$ hours.

Answer: The two cars will be 600 miles apart in 5 hours.

Problem 4: t is the time of travel for a car, $(t + 3)$ is the time of travel for a truck, $d_1 = d_2$, $65t = 50(t + 3)$, $15t = 150$, $t = 10$ hours.

Answer: The car will overtake the truck in 10 hours.

Problem 5: t is the time for a girl, $(t + 1)$ is the time for a boy. $d_1 = d_2$, $12t = 6(t + 1)$, $6t = 6$, $t = 1$ hour, $d_1 = 12t = 12(1) = 12$ miles.

Answer: The girl will travel 12 miles.

Problem 6: The distance from the starting point is d. The travel time for a bus: $t_1 = \dfrac{d}{60}$, the travel time for a car is $t_2 = \dfrac{d}{40}$, $t_1 + 2 = t_2$, $\dfrac{d}{60} + 2 = \dfrac{d}{40}$ (multiply each term by the LCD 120); $2d + 240 = 3d$, $d = 240$ miles.

Answer: The vehicles are 240 miles away from the starting point.

Problem 7: The rate of the plane is p, the rate of the wind is w. The speed of the plane with the wind is $(p + w)$, against is $(p - w)$. $d_1 = d_2 = 300$ miles. The first equation is $2(p + w) = 300$, or $p + w = 150$. The second equation is $2.5(p - w) = 300$, or $p - w = 120$. The system of two equations:

$p + w = 150$

$p - w = 120$

Solve by addition: $2p = 270$, $p = 135$ mph, substitute into $p + w = 150$, $135 + w = 150$, $w = 15$ mph.

Answer: The rate of the wind is 15 mph.

Chapter 20: The Misfits

Problem 1: Lorry's diet plan is x calories, Mary's diet plan is $2x - 950$. The equation is: $2x - 950 = 1,850$; $2x = 2,800$, $x = 1,400$.

Answer: Lorry's plan is 1,400 calories.

Problem 2: The number of hours of labor is x. Total for labor is $\$52 \cdot x$. The equation is: $52x + 450 = 1,074$, $52x = 624$, $x = 12$.

Answer: 12 hours of labor are needed.

Problem 3: The number of cars sold in December 2008 is x, the equation is: $2x + 4 = 62$, $2x = 58$, $x = 29$.

Answer: 29 cars were sold in December 2008.

Problem 4: The sum of the other two angles is $180° - 37° = 143°$.

Let x be the measure of $\angle Q$, then the measure of $\angle R$ is $143° - x$. $\dfrac{\angle Q}{\angle R} = \dfrac{x}{143 - x} = \dfrac{4}{7}$, $7x = 4(143 - x)$, $7x = 572 - 4x$, $11x = 572$, $\angle Q$ is $x = 52°$, $\angle R$ is $143° - 52° = 91°$.

Answer: $\angle Q$ is $52°$ and $\angle R$ is $91°$.

Problem 5: The initial amount of gas in a tank is x. Convert percents: $25\% = 0.25$ and $20\% = 0.20$. For the first trip, the car used $0.25x$ gallons. The rest of the tank is $x - 0.25x = 0.75x$. For the second trip, the car used $0.20(0.75x) = 0.15x$ gallons. The amount of the gas left after the two trips is $0.25x + 0.15x + 4$. The equation is: $0.25x + 0.15x + (0.25x + 0.15x + 4) = x$, $0.80x - x = -4$, $-0.20x = -4$, $x = 20$ gallons.

Answer: There were 20 gallons of gas.

Problem 6: The shortest side is x. The longest side is $2x + 5$. To find the third side, use the percent formula and find the whole b: $a = p\% \cdot b$, $x = 0.75(b)$, the third side is $\dfrac{x}{0.75}$. The equation is: $x + (2x + 5) + \dfrac{x}{0.75} = 135$. Multiply by the LCD (0.75): $0.75x + 0.75(2x + 5) + x = 0.75(135)$. Multiply by 100: $75x + 75(2x + 5) + 100x = 75(135)$, $75x + 150x + 375 + 100x = 10,125$, $325x = 9,750$, the shortest side is $x = 30$ yards, the longest side is $2x + 5 = 2(30) + 5 = 65$ yards.

Answer: The longest side is 65 yards.

Problem 7: The speed of the car is x. Solve the proportion to find the speed of the bus: $\dfrac{\text{car speed}}{\text{bus speed}} = \dfrac{x}{\text{bus speed}} = \dfrac{3}{1}$, $3 \cdot (\text{bus speed}) = 1x$, the bus speed is $\dfrac{x}{3}$. The time of travel for the car is $\dfrac{160}{x}$, the time of travel for the bus is $\dfrac{160}{\frac{x}{3}} = \dfrac{160 \cdot 3}{x} = \dfrac{480}{x}$. Convert

40 minutes into hours: 40 minutes $= \dfrac{40}{60} = \dfrac{2}{3}$. The bus traveled $2 + \dfrac{2}{3} = 2\dfrac{2}{3} = \dfrac{8}{3}$ hours

longer than the car. The equation is: $\dfrac{160}{x} = \dfrac{480}{x} - \dfrac{8}{3}$. Multiply by the LCD ($3x$) and

reduce by common factors: $480 = 1{,}440 - 8x$, $8x = 960$, $x = 120$ kmph.

Answer: The speed of the car is 120 kmph.

Index

A

absolute values, 17-19
absolute zero of temperature, 216
addition method
    phrase keywords, 20
    systems of linear equations, 43-47
addition-subtraction property of equality, 24
Additive Identity Property, 19
Additive Inverse Property, 19
age word problems, 115
    fractions, 125-129
    multiple people, 118-124
    practice problems, 129-130
    single person, 115-118
al-jabr, 12
al-Khawarizmi, 11
algebra
    history, 11-12
    phrase translation, 20-21
    rules
        absolute value, 17-19
        order of operations, 16-17
        properties, 19-20
    terminology, 12-16
    word problems, 3-4
        checking answer, 10
        crucial information, 4-6
        unit selection, 6

    variables, 8-9
    visualization, 6-8
algebrista, 12
angles, 192-197
answers, checking, 10
Archimedes, 102, 213, 231
area, 201-206
arithmetic mean, 241-246
    defined, 241-242
    practice problems, 254
    weighted average, 246-254
Associative Property of Addition, 19
Associative Property of Multiplication, 19
averages, 242-246
    practice problems, 254
    weighted, 246-254

B

balances
    level principle, 209-215
    linear equations, 23-27
base 10 system, 112

C

calculating distance, speed, and time, 257-258

Celsius
  practice problems, 220
  scale conversions, 216-219
  temperature measurement, 215-216

checking the answer, 10

coefficients, 13
  fractions, 31

coins
  counting with one equation, 149-153
  counting with two equations, 153-158
  dry mixtures, 158-161
  practice problems, 162

Commutative Property of Addition, 19

Commutative Property of Multiplication, 19

complementary angles, 192-197

complex fraction, 134

compound interest, finances, 168-172

consecutive even integers, 106

consecutive integers, 103-108

consecutive odd integers, 106

constants, combining, 13

cross-multiplication, 34

D

decimals, linear equations, 35-36

Descartes, René, 12

digit problems, 108-113

Diophantus, 11

direct proportions, 64-65
  set up, 66-68
  solving methods, 65-66

discounts
  calculating, 182-184
  double, 185-187
  original prices, 177-182

percent of, 184-185
  practice problems, 187

distance
  formula for calculating, 258
  lever principle, 209-215
  motion problems, 258
  variables, 289

Distributive Property, 19

division, phrase keywords, 21

double discounts, 185-187

Double Negative Property, 19

dry mixtures, 158-162

E

elements, motion problems, 275-278

equals signs, phrase keywords, 21

equations
  checking answers, 10
  equivalent, 24
  linear equations, 23
    balancing, 23-27
    decimals, 35-36
    fractions, 31-35
    in one variable, 24
    simplifying, 29-31
    systems, 37-47
    transforming, 27-29
    variables, 8-9

equivalent equations, 24

expanded form, 109

extended addition property of equality, 43

F

Fahrenheit
  practice problems, 220
  scale conversions, 216-219
  temperature measurement, 215-216

numbers
  irrational, 15
  natural, 14
  rational, 14
  whole, 14

O

operations, order, 16-17
opposite directions, multiple travelers, 271-272
order of operations, 16-17
ordered pairs, 37-38

P

paradoxes of motion, Zeno of Elea, 275
parallelograms, 200
per mille, 83
percents, 79
  defined, 79-80
  determining amount of change, 90-92
  discount, 184-185
  finding from whole and part, 89-90
  finding part of the whole, 81-85
  finding whole, 86-88
  practice problems, 92
perimeters, 197-201
phrases, translation, 20-21
polygons, 200
  angles, 192-197
  area, 201-206
  perimeters, 197-201
practice problems
  age problems, 129-130
  arithmetic mean, 254
  coins, 162
  discounts, 187
  finances, 176
  geometry, 206-207

lever principle, 220
liquid mixtures, 236-237
motion problems
  multiple travelers, 279
  single travelers, 265-266
multiple type problems, 284-292
number problems, 114
percents, 92
ratios, 62
temperature, 220
unidentified type, 281-284
prices
  discounts
    calculating, 182-184
    double, 185-187
    original price, 177-182
    percent of, 184-185
    practice problems, 187
  dry mixtures, 158-161
principal, 163-164
problems
  age problems, 129-130
  arithmetic mean, 254
  coins, 162
  discounts, 187
  finances, 176
  geometry, 206-207
  lever principle, 220
  liquid mixtures, 236-237
  motion problems
    multiple travelers, 279
    single travelers, 265-266
  multiple type problems, 284-292
  number problems, 114
  percents, 92
  ratios, 62
  temperature, 220
  unidentified type, 281-284
profits
  investments, 172-176
  practice problems, 176
properties, 19-20

U–V

units, word problems, 6

values, absolute, 17-19
variable x, solving linear equations, 24
variables, 12-13
    addition method, 43-47
    distance, 289
    isolating
        balancing linear equations, 23-27
        decimals, 35-36
        fractions, 31-35
        simplifying linear equations, 29-31
        transforming linear equations, 27-29
    motion problems, 269
    substitution method, 38-43
    word problems, 8-9
vertex, angles, 196
Vieta, François, 12
visualization, word problems, 6-8

W–X–Y–Z

weighted averages, 246-254
whole numbers, 14
word problems, 131-132
    algebra, 3-4
    checking answers, 10
    crucial information, 4-6
    unit selection, 6
    variables, 8-9
    visualization, 6-8

Zeno of Elea, paradoxes of motion, 275